ANCIENT HINDU GEOMETRY

THE SCIENCE OF ŚULBA

ANCIENT HINDU GEOMETRY

THE SCIENCE OF ŚULBA

By
BIBHUTIBHUSAN DATTA

COSMO
PUBLICATIONS
2021 New Delhi

ANCIENT HINDU GEOMETRY.
THE SCIENCE OF ŚULBA

ISBN-13 978-81-7020-460-2

First Published by Cosmo 1993, 2005
This Cosmo Edition 2021

Published by
COSMO PUBLICATIONS
24-B, Ansari Road, Darya Ganj,
New Delhi-110 002, India
Ph.: 011 - 4365 6579 & 011-4678 3359
Email : cosmo@cosmopublications.com
Website : www. cosmopublications.com

PRINTED AND BOUND IN INDIA

ॐ

तृप्यन्तामृषयः सर्व्वे यैर्दृष्टं शुल्वविज्ञानम् ।

व्यासो यैश्च कृतस्तस्य तेभ्यश्चेदं नमो नमः ॥

ॐ शान्तिः शान्तिः शान्तिः ॥

PREFACE

In this book an attempt has been made to study the Vedic rites of the *Agni-cayana* (or "the construction of the Fire-altar") from a point of view, purely secular, quite different from that of unravelling their deep mysticism and highly speculative philosophy. The whole purpose has been to get as much insight as possible into the knowledge and achievements of the Hindus in the science of mathematics, more particularly in its branch of geometry. The *Agni-cayana* reveals an important aspect of the Hindu genius of which the student of the Vedic culture is apt to lose sight. Most scholars, when they think of the genius of the Vedic Hindu, are naturally more attracted by his noble religion, sublime philosophy, enormous extent and most varied character of his rich literature, and charming devotional poetry. But the Vedic Hindu, in his great quest of the *Parā-vidyā* ("Supreme knowledge"), *Satyasya Satyaṃ* ("Truth of truths," "Absolute Truth"), made progress in the *Aparā-vidyā* ("inferior knowledge," "relative truths"), including the various arts and sciences, to a considerable extent, and with a completeness which is unparalleled in antiquity. Of these the special concern of this volume is with the Vedic science of geometry, technically called by the name *Śulba*.

The writer is fully conscious of his limitations to perform in the proper way the arduous task that he has undertaken. Truly he feels, to speak after the immortal poet Kālidāsa,

क्व शुल्वविहृतविद्या क्व चाल्पविषया मति।
तितीर्षुरुडुपेनापि दुस्तरमस्मि सागरम्॥

" How great is the science which revealed itself in the *Śulba*, and how meagre is my intellect! I have aspired to cross the unconquerable ocean in a mere raft." Moreover, the work had had to be done hurriedly within a short time at his disposal just on the eve of his retirement from active life in the University, in 1930, amidst other preparatory arrangements consequent thereto. So it could not be made as comprehensive and thorough as it should have been. It is nevertheless the author's confident hope that this imperfect sketch will create a lively interest in the early Hindu geometry amongst the historians of mathematical sciences.

It is a pleasure to express indebtedness to my teacher, Professor Ganesh Prasad, for his interest and encouragement for the work. In deference to his wish, I delivered, by special invitation of the authorities, a course of six lectures on the science of the *Śulba*, in the University of Calcutta, during December, 1931. I tender grateful thanks to Mr. Atul Chandra Ghatak, Superintendent, and the staff of the Calcutta University Press for kindly expediting the book through the Press in order to help me to go back to my retirement earlier. Above all, I remember with pleasure the name of my younger brother, Dr. Binode Behari Datta, M.A., Ph.D., for his help and association in every way in this book.

CONTENTS

CHAPTER I

Śulba

CHAPTER II

Commentators

CHAPTER III

Growth and Development of the Śulba

CHAPTER IV

Postulates

CHAPTER V

Constructions

CHAPTER VI

COMBINATION OF AREAS

CHAPTER VII

TRANSFORMATION OF AREAS

CHAPTER VIII

AREAS AND VOLUMES

CHAPTER IX

THE THEOREM OF THE SQUARE OF THE DIAGONAL

CHAPTER X

RATIONAL RECTANGLES

CHAPTER XI

SQUARING THE CIRCLE

CHAPTER XII

SIMILAR FIGURES

CHAPTER XIII

GEOMETRICAL ALGEBRA

CHAPTER XIV

INDETERMINATE PROBLEMS

CHAPTER XV

ELEMENTARY TREATMENT OF SURDS

CHAPTER XVI

FRACTIONS AND OTHER MINOR MATTERS

ABBREVIATIONS

ĀpŚl=*Āpastamba Śulba.*
ĀpŚr=*Āpastamba Śrauta.*
BŚl=*Baudhāyana Śulba.*
BŚr=*Baudhāyana Śrauta.*
KapS=*Kapiṣṭhala Saṁhitā.*
KŚl=*Kātyāyana Śulba.*
KŚlP=*Kātyāyana Śulba Pariśiṣṭa.*
KŚr=*Kātyāyana Śrauta.*
KṭS=*Kāṭhaka Saṁhitā.*
MaiS=*Maitrāyanīya Saṁhitā.*
MaiŚl=*Maitrāyanīya Śulba.*
MāŚl=*Mānava Śulba.*
MāŚr=*Mānava Śrauta.*
PañBr=*Pañcaviṁśa Brāhmaṇa.*
ṚV=*Ṛg-veda.*
SBE=Sacred Books of the East Series.
ŚBr=*Śatapatha Brāhmaṇa.*
TBr=*Taittirīya Brāhmaṇa.*
TS=*Taittirīya Saṁhitā.*
VS=*Vājasaneya Saṁhitā.*
ZDMG=Zeitschrift der deutschen morgenländischen Gesellschaft.

CHAPTER I

Sulbas

The *Sulbas*, or as they are more commonly known at present amongst oriental scholars, the *Sulba-sūtras*, are manuals for the construction of altars which are necessary in connexion with the sacrifices of the Vedic Hindus. They are sections of the *Kalpa-sūtras*, more particularly[1] of the *Srauta-sūtras*, which form one of the six *Vedāṅgas* (or "The Members of the Veda") and deal specially with rituals or ceremonials. Each *Srauta-sūtra* seems to have its own *Sulba* section. So there were, very likely, several such works in ancient times.[2] At present we know, however, of only seven *Sulba-sūtras*, those belonging to the *Srauta-sūtra* of Baudhāyana, Āpastamba, Kātyāyana,

[1] The *Kalpa-sūtras* are broadly divided into two classes, the *Gṛhya-sūtras* (or "The rules for ceremonies relating to family or domestic affairs" such as marriage, birth, etc.) and the *Srauta-sūtras* ("The rules for ceremonies ordained by the Veda" such as the preservation of sacred fires, performance of the sacrifices, etc.). The *Sulba-sūtras* belong to this latter class.

[2] We have it on the authority of Patañjali (150 B.C.), the Great Commentator of Paṇini's Grammar, that there were as many as 1,131 or 1,137 different schools of the *Veda*.

"एकविंशतिधा वा बाह्वृच्यम्, एकशतमध्वर्युशाखाः ।
सहस्रवर्त्मा सामवेदः, नवधा आथर्वणो वेदः पञ्चदशभेदो वा ॥"

Or "There were 21 different schools of the *Ṛg-veda*; 101 schools of the *Yajur-veda*; 1,000 of the *Sāma-veda*; and 9 or 15 of the *Atharva-veda*." Each school of the *Veda* had its own *Śrauta-sūtra* and hence probably its own *Sulba*. Thus it seems that there were numerous manuals of geometry in ancient India. But most of them are now lost.

Mānava, Maitrāyaṇa, Vārāha and Vādhula.[1] These manuals are also found separately.

As related to the different *Vedas*, the *Śulba-sūtras* of Baudhāyana, Āpastamba, Mānava, Maitrāyaṇa and Vārāha belong to the *Kṛṣṇa Yajur-veda*; and the *Kātyāyana Śulba-sūtra* to the *Śukla Yajur-veda*.

It was perhaps primarily in connexion with the construction of the sacrificial altars of proper size and shape that the problems of geometry and also of arithmetic and algebra presented themselves, and were studied in ancient India, just as the study of astronomy is known to have begun and developed out of the necessity for fixing the proper time for the sacrifice.[2] At any rate, from the *Śulba-sūtras*, we get a glimpse of the knowledge of geometry that the Vedic Hindus had.[3] Incidentally they furnish us with a few other subjects of much mathematical interest.

Of all the extant *Śulbas*, that of the Baudhāyana is the biggest and is also, perhaps, the oldest. It is divided into three chapters. The first chapter contains 116 *sūtras* ("aphorisms") of which the opening two are merely introductory; *sūtras* 3-21 define the various measures ordinarily employed in the *Śulbas*; *sūtras* 22-62 give the more important of the geometrical propositions necessary for the construction of the sacrificial altars; and *sūtras* 63-116 deal briefly with the relative positions

[1] In the commentary of Karavindasvāmī on the *Āpastamba Śulba* (xi. 11), we find reference to two other works, *viz.*, *Maśaka Śulba* and *Hiraṇyakeśi Śulba*, which are not available now. There is also a quotation from the latter work (*ĀpSl*, vi. 10).

[2] Bibhutibhushan Datta, "The Scope and Development of the Hindu *Gaṇita*," *Ind. Hist. Quart.*, Vol. 5 (1929), pp. 479-512.

[3] There are reasons to believe that side by side with the practical geometry of the *Śulbas*, the Vedic sacrificial priests had also an esoteric geometry as their secret property.

and spatial magnitudes of the various *vedis* (or "altars"). The second chapter consists of 86 *sūtras* of which the major portion, *sūtras* 1-61, is devoted to the description of the spatial relations in the different constructions of the *Agnis* (or "the large Fire-altars made of bricks") in general, and the remaining portion, *sūtras* 62-86, elaborates the construction of the two simplest Agnis, *viz.*, the *Gārhapatya-citi* (or "The House-holder's Fire-altar") and *Chandaś-citi*[1] (or "The Agni made, as it were, of *mantras* instead of bricks"). The third chapter, in altogether 323 *sūtras*, describes the construction of as many as seventeen different kinds of *Kāmya Agnis* (or "the altars for the sacrifices performed with a view to attain definite objects") of rather complex nature. In case of some, the description is quite elaborate and minute in details, but in other cases it is less so.

The *Śulba-sūtra* of Āpastamba is broadly divided into six *paṭalas* (or "sections"). Of these the first, third and the fifth are each subdivided again into three *adhyāyas* (or "chapters") and each of the remaining sections into four chapters. So that altogether the work contains twenty-one chapters and 223 *sūtras*. The first section of the manual, chapters i-iii, gives the important geometrical propositions required for the construction of altars. The second section or the chapters iv-vii, describe the relative positions of the various *vedis* and their spatial magnitudes. Unlike Baudhāyana, Āpastamba here indicates

[1] In case of the *Chandaścıti*, the *agnicit* ("the Fire-altar-builder") draws on the ground the Agni of the prescribed shape, ordinarily of the primitive shape of the falcon. He then goes through the whole prescribed process of construction imagining all the while as if he is placing every brick in its proper place with the appropriate *mantras*. The *mantras* are, indeed, muttered but the bricks are not actually laid. Hence the name *Chandaściti*, that is, the *citi* or altar made up of *chandas* or Vedic *mantras* instead of bricks or loose mud pieces.

briefly also the methods of their construction. They are of course the particular applications of the general geometrical theorems taught in the earlier section. The remaining sections of the *Āpastamba Śulba-sūtra*, comprising the chapters viii-xxi deal with the construction of the *Kāmya Agnis*. It is noteworthy that almost the same set of geometrical propositions are taught by both Baudhāyana and Āpastamba. But the latter has treated of a smaller number of varieties of the *Kāmyas* than the former. For instance, Āpastamba teaches only one kind of *ratha-cakra-citi* (or "the wheel-shaped altar") whereas Baudhāyana gives two.

The *Śulba-sūtra* of Kātyāyana, also known as *Kātyāyana Śulba-pariśiṣṭa* or *Kātīya Śulba-pariśiṣṭa*, is divided into two parts. The first part is composed in the style of the *sūtras* or aphorisms, as those noted above, while the second part is composed in verses. The earlier part is again subdivided into seven *kaṇḍikās* (or "short sections") containing altogether 90 *sūtras*. It teaches the geometrical propositions, the different measures employed in the work, and the relative positions and spatial relations for the different constructions of the *Agnis*. This manual does not treat of the construction of the *Kāmya Agnis*. It is because that subject has been treated in a different chapter of the *Kātyāyana Śrauta-sūtra*.[1] The second part comprises nearly about 40 or 48 verses.[2] It gives mainly a description of the measuring tape (*rajju*),

[1] *KŚr*, Chap. xvii.

[2] There is a bit of uncertainty about the total number of verses in the *Pariśiṣṭa* of the *Kātyāyana Śulba*. The manuscript of it that is preserved in the Library of the India Office, London (No. E 363), has 48 verses, whereas the manuscript in possession of the Bhandarkar Institute, Poona (No. 74 of A 1881-82), shows only 40 verses. The latter MS. also includes the commentary of Mahīdhara on that manual and he counts 43 verses.

the gnomon, the attributes of an expert altar-builder and also a few general rules for his conduct. Some of the processes of construction described in the earlier part together with a few other new matters, though of comparatively minor importance, also appear there. I think the title *Kātyāyana Śulba-pariśiṣṭa* or ("The Appendix to the *Śulba* of Kātyāyana") was originally designed for this part and should be kept reserved to it, even now. For it is really a sort of an appendix to the earlier part, the *Kātyāyana Śulba* proper. The commentator Rāma is also of the same opinion as we are. And the same differentiation is found to have been scrupulously maintained by Yājñika Deva, the commentator of the *Kātyāyana Śrauta-sūtra.* Kātyāyana observes that the second part, especially the recapitulations in it, was meant to help those whose intellects are too poor to be able to fully grasp the inner meanings of the compositions in the sūtra style. Compared with the works of Baudhāyana and Āpastamba, the *Śulba* of Kātyāyana presents some interesting features as it exhibits the whole body of geometrical knowledge required for the Vedic altar-builder in a more systematic form.

The *Śulba-sūtra* of Manu is a small treatise composed in both prose and verse. It is divided into seven *khaṇḍas* (or "parts," "sections"). In the first section is given a description of the measuring tape, the gnomon, measures, four methods of determining the cardinal directions and also a method of constructing a square on a given straight line. It may be noted that we do not find in the Āpastamba and Baudhāyana *Śulba-sūtras* any method of determining the cardinal directions, though it is essentially necessary for the proper construction of the sacrificial altars to have an accurate knowledge about them. They proceed on the assumption that the cardinal directions are already known. Kātyāyana teaches three methods for the same, while Manu teaches as many as four. The sections ii-vi treat of the relative positions, spatial magnitudes and also the methods of the construction of the the different *vedis*. Here we find mention of certain *vedis*, *e.g.*, the *Pākayājñikī*, *Māruti* and *Vāruṇī vedis* which are not included in the abovementioned manuals. The last section of the *Mānava Śulba-sūtra* furnishes us with some hints about the sacrificial fees. It also describes the method of the construction of the *Suparṇa-citi*. This *citi* is not found in other *Śulba-sūtras*. But for the head, its spatial magnitudes are the same as those of the most primitive *citi*, the *Saptavidha-sāratni-prādeśa-caturasra-śyena-cit*, described by Baudhāyana and others.

The *Maitrāyaṇīya Śulba-sūtra* is a different recension of the *Mānava Śulba-sūtra*. They cover almost the same ground and, more than that, many passages of them are identical. But still they should not be mistaken as one and the same work. The arrangement of matter in them is not parallel. And there are also other marks of distinction between them. The *Maitrāyaṇīya Śulba-sūtra* is comprised of four *khaṇḍas* (or "sections").

The *Vārāha Śulba-sūtra* is very closely related to the above two works. There are found several repetitions between these works. This will not seem strange if we remember that they belong to the same school of the *Kṛṣṇa Yajur-veda*. Similarly we find in these *Śulba-sūtras* repetition of a few verses of the *Kātyāyana Śulba-pariśiṣṭa*. The *Vārāha Śulba-sūtra* is broadly divided into three parts and each part is again subdivided into several sections.

As regards their importance, the available *Śulba-sūtras* can sharply be divided into two classes. The first class will include the manuals of Baudhāyana, Āpastamba and Kātyāyana. They give us an insight into the early state of Hindu geometry before the rise and advent of the Jaina Sect (500-300 B.C.).[1] The *Śulba-sūtras* of Mānava, Vārāha, Maitrāyaṇa and Vādhula add practically very little to our stock of information in this respect. So they may be considered to be of minor importance from our point of view.

In the title *Śulba-sūtra*, the word *sūtra* means an "aphorism," "a short rule." It simply describes the style of the composition of the works and has practically no reference to their subject-matter. The science itself is really called the *Śulba*. And that is, in fact, the original title of the manuals. It is by this title that the *Śulba-sūtra* of Āpastamba has been mentioned in his *Śrauta-sūtra*.[2] The commentators are oftentimes found to speak of the *Śulba* of Baudhāyana, the *Śulba* of Āpastamba, etc. This will be further confirmed by the commonly known title of the second part of the work attributed to Kātyāyana, namely the *Śulba-pariśiṣṭa* (or "The Appendix to the *Śulba*") and also by the title *Śulbī-kriyā* (or "The Practice of the *Śulba*") given to that appendix in itself. Thus it is proved conclusively that the true name of the subject is *Śulba*. As the *Śulba* deals with the science of geometry and its application as known amongst the early Hindus, we conclude that the earliest Hindu name for geometry was *Śulba*. Geometry was then sometimes also called *Rajju*, as is evident from the opening sūtra of the *Śulba* of Kātyāyana, "I shall speak of the 'Collection of (rules regarding) the *Rajju*.'" There are many other reliable pieces of evidence leading strongly to the same conclusion.[3]

1 For an insight into Hindu geometry after the advent of the Jainas the reader is referred to the author's article, "Geometry in the Jaina Cosmography," in *Quellen und Studien zur Geschichte der Mathematik*, Abteilung B, Bd. 1, 1930, pp. 245-254.

2 *ĀpŚr*, xvii. 26. 2.

3 Bibhutibhushan Datta, "Origin and History of the Hindu Names

In Sanskrit, the words *śulba* and *rajju* have the identical significance, which is ordinarily "a rope," "a cord." The word *śulba* or *śulva* is derived from the root *śulb* or *śulv* meaning "to measure" and hence its etymological significance is "measuring" or "act of measurement." From that it came to denote "a thing measured" and consequently "a line (or surface)" as well as "an instrument of measurement" or "the unit of measure." Thus the terms *śulba* or *rajju* have four meanings: (1) mensuration—the act and process of measuring; (2) line (or surface)—the result obtained by measuring; (3) a measure—the instrument of measuring; and (4) geometry—the art of measuring. In the ancient literature of the Hindus we indeed find mention of three kinds of measure—linear, superficial as well as voluminal—having the same epithet *rajju*. In the *Śulbas* the measuring tape is called *rajju*. And we further find there the use of the word in the sense of "a line" also. For instance, we have the term *akṣṇayā-rajju*="diagonal line." Kātyāyana observes:[1] "(The terms) *karaṇī* ('producer'), *tat-karaṇī* [2] ('that-producer'), *tiryaṅmānī* ('transverse measurer'), *pārśvamānī* ('side measurer'), and *akṣṇayā* ('diagonal') *rajjus* are ('lines')."

In the *Mānava Śulba*[3] and *Maitrāyaṇīya Śulba*,[4] the science of geometry is called the *Śulba-vijñāna* (or "the Science of the *Śulba*").[5] One who was well versed in that

for Geometry," *Quellen und Studien z. Gesch. d. Math.*, Abteil. B, Bd. 1, 1930, pp. 113-9.

[1] *KŚl*, ii. 7.

[2] That is, *dvi-karaṇī*, *tri-karaṇī*, etc.

[3] *MāŚl*, iii. 2.

[4] *MaiŚl*, Ch. i.

[5] This term and also the terms *śulba-vid* and *śulba-paripṛcchaka* for an expert in the *Śulba*, will further support our conclusion as regards the earliest Hindu name for geometry.

science was called in ancient India as *saṁkhyājña* (or "the expert in Numbers"), *parimāṇajña* ("the expert in measuring"), *sama-sūtra-nirañchaka* ("uniform-rope-stretcher"), *Śulba-vid* ("the expert in the *Śulba*") and *Śulba-paripṛcchaka* ("the inquirer into the *Śulba*").[1] Of these, one term, *viz.*, *sama-sūtra-nirañchaka*, perhaps deserves more particular notice. For we find an almost identical term, *harpedonaptae* ("rope-stretcher"), appearing in the writings of the Greek Democritos (*c.* 440 B.C.). It seems to be an instance of Hindu influence on Greek geometry. For the idea in that Greek term is neither of the Greeks nor of their acknowledged teachers in the science of geometry, the Egyptians, but it is characteristically of Hindu origin.) In the Pāli literature, we find the terms *rajjuka* and *rajju-grāhaka* ("rope-holder") for the king's land surveyor.[2] The first of these terms appears copiously, in its various case-endings, in the inscriptions of the Emperor Aśoka (250 B.C.). In the comparatively later *Śilpa-śāstras*, the surveyor is spoken of as *sūtra-grāhī* or *sūtra-dhāra* ("rope-holder") and he is further described as an expert in alignment (*rekhā-jña*, lit. "one who knows the line").

CHAPTER II

Commentators

There are now available several commentaries on the *Śulbas*. The more important manuals are found to have been commented upon by more than one writer. Thus we have two commentaries on the *Śulba* of Baudhāyana. One of them is by Dvārakanātha Yajvā and is named *Śulba-dīpikā* (" The Light of the *Śulba* "). The other, called *Śulba-mīmāṁsā* (" The investigation into the *Śulba* "), is by Veṅkateśvara Dīkṣita. On the *Āpastamba Śulba*, there are as many as four well-known commentaries: (1) *Śulba-vyākhyā* ("The Explanation of the *Śulba* ") by Kapardisvāmī, (2) *Śulba-pradīpikā* (" The Light of the *Śulba* ") by Karavindasvāmī, (3) *Śulba-pradīpa* (" The Light of the *Śulba* ") by Sundararāja and (4) *Āpastambīya Śulba-bhāṣya* (" The commentary on the *Śulba* of Āpastamba ") by Gopāla, son of Gārgya Nṛsiṁha Somasuta. Sundararāja's work is also called *Sundara-rājīya* (" The work of Sundararāja ") after the name of the author, as is usual in Sanskrit. I have come across two commentaries on the *Kātyāyana Śulba*, namely, *Śulba-sūtra-vṛtti* (" The Explanation of the *Śulba-sūtra* ") of Rāma or Rāmacandra, son of Sūryadāsa and *Śulba-sūtra-vivaraṇa* (" The Exposition of the Śulba-sūtra ") by Mahīdhara.

The dates of most of the commentators of the *Śulbas*, more particularly of the notable ones, have not as yet been ascertained, even approximately. Nor is it easy to do so. The periods to which some of them can be assigned from the reference by them to anterior writers and from the reference to them by writers posterior lie within such widely varying limits as to be of no tangible value. We shall begin here with the notice of those commentators whose times are known either definitely or very nearly so.

We find from the colophon that Mahīdhara completed his commentary on the *Kātyāyana Śulba* in the Saṁvat year 1646 (=1589 A. D.), at Benares. It is also stated there that that commentary is based on the *Śulba-sūtra-vṛtti* of Rāma. Mahīdhara wrote as many as seventeen works on various subjects. His *Mantramahodadhi* was completed in 1589 A.D. and *Viṣṇubhakti-Kalpalatā-prakāśa* in 1597.

The commentator Rāma was an inhabitant of Naimiṣa (near modern Lucknow). He seems to have been the author of several works such as *Karma-dīpikā*, *Kuṇḍākṛti* (with commentary), *Śulba-vārttika*, *Śāṅkhyāyana Gṛhya-paddhati*, *Samara-sāra* and its commentary, *Samara-sāra-saṁgraha* and the commentaries on the *Kātyāyana Śulba* and *Śāradā-tilaka Tantra*. The date of composition of the *Kuṇḍākṛti* is given as 1506 Vikrama Saṁvat (1449 A.D.). In his commentary on the *Kātyāyana Śulba*, Rāma has quoted copiously from his *Śulba-vārttika* ("The Critical Annotation of the *Śulba* ") and also from his commentary on the *Śāradā-tilaka*. There is also a quotation from the *Triśatikā* of Śrīdhara (c. 750 A.D.).[1] In this work we notice some new contributions from him. To construct a right-angled triangle having a given leg (a), Rāma suggests the employment of a new rational rectangle (a, $8a/15$, $17a/15$) in addition to those taught in the *Śulbas*.[2] But the most notable contribution of him is a correction to the well-known *Śulba* value of $\sqrt{2}$, *viz.*,

$$\sqrt{2}=1+\frac{1}{3}+\frac{1}{3.4}-\frac{1}{3.4.34} \qquad (1)$$

[1] *KŚl*, i. 30 (*com.*). The quoted passage is the Rule 47 of the *Triśatikā* of Śrīdhara, but there is no mention of any name.

[2] *KŚl*, i. 15 (*com.*).

Rāma shows that a more accurate (*sūkṣmatara*) value will be given by [1]

$$\sqrt{2}=1+\frac{1}{3}+\frac{1}{3.4}-\frac{1}{3.4.34}-\frac{1}{3.4.34.33}+\frac{1}{3.4.34.34}. \quad (2)$$

Turned into decimal fraction, the expression (1) gives $\sqrt{2}=1{\cdot}4142156863...$and (2) yields $\sqrt{2}=1{\cdot}414213502......$ According to modern calculation, $\sqrt{2}=1{\cdot}41421356...$ So that Rāma's value for $\sqrt{2}$ is correct up to seven places of decimals whereas the *Śulba* value is up to five places.

It should perhaps be noted that for certain inconsistency with the *Śrauta-sūtra* of Kātyāyana, Rāma suspects that the *Śulba* attributed to Kātyāyana might be written by a different person.[2] But the inconsistency is so minor that we cannot subscribe to the opinion of Rāma in this matter. It can be reasonably explained in other ways.

Śivadāsa, son of Nārada, a resident of the city of Benares, wrote a commentary on the *Mānava Śulba.* His younger brother, Śaṅkarabhaṭṭa is the commentator of *Maitrāyaṇīya Śulba.* Both the brothers quote from Rāma Bājapeya, who is no other than the commentator of the *Kātyāyana Śulba.* Śivadāsa has quoted the second Bhāskara's (1150 A.D.) *Rule of Three*, and also from his *Līlāvati*, by name. He must have been posterior to the celebrated Sāyana (1320-1380 A.D.) whom he quotes.

Śivadāsa observes :

" The study of the *Śulba* should be begun after having finished the study of the science of mathematics; otherwise there cannot be a thorough knowledge of the *Śulba.*"

1 *Ibid*, ii. 13 (*com.*). The rationale of this is stated to have been given in the *Śulba-vārttika* whence this and the preceding matters have been taken.

2 Compare *KŚl*, ii. 8 (*com.*).

Of the known commentaries of the *Āpastamba Śulba* the earliest one is, I have good reasons to believe, that of Kapardisvāmī. This writer is known to have commented also on *Āpastamba-Śrauta-sūtra*, *Āpastamba-sūtra-paribhāṣā*, *Darśapaurṇamāsa-sūtra*, *Bhāradvāja Gṛhya-sūtra*, etc. He is quoted by Śūlapāṇi, Hemādri, Nīlakaṇṭha and others. Now Śūlapāṇi lived near about 1150 A.D. He was the teacher of the famous Ṣaḍguruśiṣya (1143-1193 A.D.), the author of the *Vedārthadīpikā*. Hemādri was the minister of King Mahādeva (1260-71) of Devagiri and of his nephew and successor Rāmachandra (1271-1309). So Kapardisvāmī lived before the twelfth century of the Christian era. He has generalised a method taught in the *Śulbas* for finding the rational right-angled triangles having a given leg. He says:

"The added portion is divided into as many parts as the number obtained by dividing the (given) leg with the added portion by half the added portion; (put) the *nirañchana* mark by diminishing the added portion by one part."

Let a be the given leg and suppose it to be increased by adding a portion a/m, where m is any rational integer. Dividing the increased length by half the increment, we get

$$\left(a+\frac{a}{m}\right)\div\frac{a}{2m}=2(m+1).$$

So that the added portion a/m shall have to be divided into $2(m+1)$ parts. Then the *nirañchana* mark is to be made at a distance

$$\frac{a}{m}-\frac{a}{m}\div 2(m+1)=\frac{a}{m}-\frac{a}{2m(m+1)}=\frac{(2m+1)a}{2m(m+1)}.$$

So that

$$a^2+\left(\frac{2m+1}{2m^2+2m}\right)^2a^2=\left(\frac{2m^2+2m+1}{2m^2+2m}\right)^2a^2.$$

This method is equally available even when m is a rational fraction; that is, when the given leg is increased also by a multiple of it, instead of by only a sub-multiple of it. But this further generalisation seems to have escaped the notice of Kapardisvāmī. At any rate, his statement does not expressly show that he meant both the cases by his generalisation. Karavindasvāmī is, however, very explicit to leave no doubt in our mind in this respect.

He says: [1]

" In case of all additions, as many times the added portion as the sum of the given side and the added portion is, into twice so many parts the added portion is divided; make the mark there (*i.e.*, in the added portion, at a distance) less by one such part. For instance, in case of adding to the given side its half, (consider) that half as one part; the given side contains two such parts. So the given side with its increment contains three parts like the increment. Dividing the added portion into twice as many parts, that mark will be (at a distance) less by one-sixth the added portion. So in case of increasing the given side by itself, the increment is one part ; the given side has one part like it. So the given side with its increment has two parts. On dividing the increment into twice as many parts, it will be divided into four parts; then the mark will be (at a distance) less by the fourth part. Similarly in case of adding the third part, the added portion is one part; the given side contains three such parts. On dividing the added portion into twice as many parts, the mark will be (at a distance) less by its one-eighth part. In the same way in case of adding the fourth and other parts, the sum of the given side and its increment should be divided into parts in the same way and the mark should be made (at a distance) less by one such

[1] *ĀpŚl*, i. 2 (*com.*).

part. Now, when the increment happens to be equal to the given side, how is then the given side to be divided? How also the mark (should be made)? How also in the case when the increment happens to be greater (than the given side)? There also the method is exactly the same, we say. But the given side with its increment should then be reduced to common denominators; the added portion should then be divided into twice the number of parts (thus obtained) and the mark should be made in it (at a distance) less by one such part. For instance, in case of adding twice as much, the increment is one part and (in terms of it) the given side is a half part. Then on adding together the increment and the given side after reduction to common denominators, there will be three halves. On dividing the increment into twice that number of parts, there will be six halves in the denominator; so the mark will be (at a distance) less by one of these parts. In case of adding three times, the increment is one part; the given side is the third part (of that); there the increment contains three third parts. So the given side and the increment together contain four third parts. On dividing the increment into twice as many parts there will be eight third parts in the denominator. Then the mark will be (at a distance) less by one of these parts. In the cases of adding four times, etc., the divisions and the marks should be made in the same way."

Let a be the given side; let it be increased by its mth part.

$$\left(a+\frac{a}{m}\right)\div\frac{a}{m}=m+1$$

Then $$\frac{a}{m}\div 2(m+1)=\frac{a}{2m(m+1)}$$

$$a+\frac{a}{2m(m+1)}=\left(\frac{2m^2+2m+1}{2m^2+2m}\right)a$$

$$\frac{a}{m}-\frac{a}{2m(m+1)}=\left(\frac{2m+1}{2m^2+2m}\right)a$$

So it follows

$$a^2+\left(\frac{2m+1}{2m^2+2m}\right)^2a^2=\left(\frac{2m^2+2m+1}{2m^2+2m}\right)^2a^2 \qquad (1)$$

Or let the given side a be increased n times it.

$$(a+na)\div na=\frac{n+1}{n}$$

$$na\div 2\left(\frac{n+1}{n}\right)=\frac{n^2a}{2n+2}$$

$$a+\frac{n^2a}{2n+2}=\left(\frac{n^2+2n+2}{2n+2}\right)a$$

$$na-\frac{n^2a}{2n+2}=\left(\frac{n^2+2n}{2n+2}\right)a$$

So it follows

$$a^2+\left(\frac{n^2+2n}{2n+2}\right)^2a^2=\left(\frac{n^2+2n+2}{2n+2}\right)^2a^2 \qquad (2)$$

The two results can be combined into one

$$a^2+\left(\frac{r^2+2r}{2r+2}\right)^2a^2=\left(\frac{r^2+2r+2}{2r+2}\right)^2a^2$$

where r is any rational number integral or fractional.

Karavindasvāmī, indeed, wrote a commentary on the whole of the *Śrauta-sūtra* of Āpastamba. He is known to be the author of a few other works also. His time is still very uncertain. He is found to have quoted, without any mention of name, certain passages from the *Āryabhaṭīya* (499 A.D.) of Āryabhaṭa I (born 476).[1] So he

[1] *ĀpŚl*, iii. 5 (com.); the reference is to the *Āryabhaṭīya*, ii. 9.

undoubtedly flourished after the fifth century of the Christian era. Though we are not in a position to fix or even suggest any closer upper limit to his time, this limit seems to us to be too earlier. There is, however, one passage in his commentary on the *Āpastamba Śulba* which might lead one to take him as belonging to a very early age. He has referred to a certain treatise on mathematics which gives an incorrect formula for the calculation of the area of a segment of a circle:[1]

$$\text{Area of a segment} = \frac{\text{arc}}{2} \times \frac{\text{arrow}}{2}.$$

This formula is not found in any known treatise on Hindu mathematics and we further know that from the time of Śrīdhara (*c.* 750), the Hindu mathematicians used a more approximate formula for the calculation of the area of the segment of a circle. Does it then follow that Karavindasvāmī lived in an age before the time of the discovery of that formula, that is, before 750 A.D.? It may be noted that the other formulæ in connexion with the mensuration of the segment of a circle have been stated as correctly as we find in the works of Brahmagupta (628) and other early Hindu mathematicians.

We are equally uncertain about the time of Sundararāja. This much we are sure that he lived before the fourth quarter of the sixteenth century of the Christian era. For it appears from the post-colophon that the copy of the manuscript of his commentary on the *Āpastamba Śulba* now in the possession of the State Library of

1 "शरार्द्धं प्रमाणेन धनुरर्द्धमभ्यस्य धनुषो फलावगम इत्यादि गणितशास्त्रादवगन्तव्यमिति ।"—*ĀpŚl*, iii. 5 (*com.*).

But on a different occasion (*ĀpŚl*, vii. 14-15, *com.*), he says

"शराहतं तु कोदन्डदलितो धनुषः फलम् ।"

This indeed gives accurately the area of the sector of the circle.

Tanjore (No. 9160) was made in Saṁvat 1638 (=1581 A.D.), and that in the Government Collection of the Asiatic Society of Bengal in Saṁvat 1645 (=1588 A.D.). Sundararāja is found to have quoted from Dvārakanātha Yajvā's commentary on the *Baudhāyana Śulba* a few passages dealing with the transformation of a square into a rectangle having a given side, the correction to the Śulba formula for squaring the circle and *vice versa*, enlargement of an altar and certain other matters.

Dvārakanātha must be posterior to Āryabhaṭa I (499) whom he quotes.[1] He proves with the help of illustrative examples that the methods taught in the *Śulbas* for the squaring of a circle and *vice versa* do not lead to an accurate result as compared with that obtained by the method of Āryabhaṭa. If $2a$ be the side of the square equivalent to the circle of radius r, then according to the *Śulba*,

$$r = a + \frac{a}{3}(\sqrt{2}-1),$$

$$a = r - \frac{r}{8} + \frac{r}{8.29} - \frac{r}{8.29.6} + \frac{r}{8.29.6.8}.$$

These lead to $\pi = 3{\cdot}0883\ldots$, $3{\cdot}0885\ldots$, respectively. Dvārakanātha Yajvā emends them to [2]

$$r = \left\{ a + \frac{a}{3}(\sqrt{2}-1) \right\} \left(1 - \frac{1}{118} \right),$$

$$a = \left(r - \frac{r}{8} + \frac{r}{8.29} - \frac{r}{8.29.6} + \frac{r}{8.29.6.8} \right) \left(1 + \frac{1}{2}.\frac{3}{133} \right).$$

These will work out $\pi = 3{\cdot}141109\ldots$, $3{\cdot}157991\ldots$.

1 *BŚl*, i. 60 (*com.*); the quoted passages are *Āryabhaṭīya*, ii. 7, 10.

2 *BŚl*, i. 60 (*com.*).

Dvārakanātha Yajvā states that

$$\frac{10}{\sqrt{3}}\ \textit{aṅgulis} = 5\ \textit{aṅgulis}\ 27\tfrac{1}{2}\ \textit{tilas},$$

$$\frac{12}{\sqrt{3}}\ \textit{aṅgulis} = 6\ \textit{aṅgulis}\ 32\ \textit{tilas},$$

$$\frac{15}{\sqrt{3}}\ \textit{aṅgulis} = 8\ \textit{aṅgulis}\ 23\ \textit{tilas}.$$

From these we get

$$\frac{1}{\sqrt{3}} = \cdot 580\ldots,\ \cdot 5784313725\ldots,\ \cdot 5784313729\ldots$$

According to modern calculation $1/\sqrt{3} = \cdot 5773\ldots$

It should be noted that there were also other commentators of the *Śulbas* anterior to those who are known to us now. Kapardisvāmī, the earliest known commentator of the *Āpastamba Śulba* has referred to at least one such anterior commentator.[1]

1 "केचिदत्र सहशब्दश्रवणादतिशयत्तृतीयेनेति वर्णयन्ति चैवसमत्वाय, तदनिष्पितमिति चाचार्यस्याङ्गव्यायामेनेति वदत... ।" *ĀpŚl*, vii. 10 (*com.*).

CHAPTER III

GROWTH AND DEVELOPMENT OF THE ŚULBA

It has already been observed that the science of geometry originated in India in connexion with the construction of the altars for the Vedic sacrifices. We now propose to treat this point more fully. We shall further trace, as far as possible, the growth and development of the Hindu Geometry from its earliest state down to the one in which we find it now in the *Śulba*. Much has been done before in this respect, by Bürk in his masterly introduction to his edition of the *Āpastamba Śulba*.[1] Much more still remains to be done.

The Vedic sacrifices are mainly of two classes: *Nitya* (or "indispensable," "obligatory") and *Kāmya* ("optional," "intentional"). The performance of the sacrifices of the former class is obligatory upon every Vedic Hindu. It will be a sin for him if he does not do them. But it is not so with the sacrifices of the second kind. For they are to be performed each with the sole motive of achieving a special object. Those who do not aim at the attainment of any such object need not perform any of them.

According to the strict injunctions of the Hindu *Śāstra* (or "Holy Scriptures") each sacrifice must be made in an altar of prescribed shape and size. It is stated that even a slight irregularity and variation in the form and size of the altar will nullify the object of the whole ritual and may even lead to an adverse effect. So the greatest care has to be taken to have the right shape and size of the altar.

[1] *Āpastamba-Śulba-sūtra*, edited and translated with an introduction by Albert Bürk, *ZDMG*, LV and LVI.

There are multitudes of the altars Of the *Nitya Agni* (or "the altars for the obligatory sacrifices"), the three primary ones are the *Gārhapatya*, *Āhavanīya* and *Dakṣiṇa*. Every Vedic Hindu has to offer sacrifices in them daily. Other obligatory sacrifices are seasonal and are performed at special periods. According to the nature of the oblations they are broadly subdivided into three groups: (1) *Iṣṭi Yajña* (or "sacrifice with oblations of butter, fruit, etc.") such as *Darśa* and *Paurṇamāsa* sacrifices which are performed at every new-moon and full-moon respectively; (2) *Paśu Yajña* (or "Animal sacrifice") such as *Nirūḍhapaśubandha* which must be performed once every year, more particularly, on a new-moon or full-moon day in the rainy season; or according to a different school twice every year at the time of the winter and summer solstices; (3) *Soma Yajña* ("Soma sacrifice"). This last sacrifice is very big and expensive and so cannot be performed often. But it must be performed in a family of Vedic Hindus at least once in three generations.

Now we find it from the *Sulba*, that the altar of the *Gārhapatya* must be of the form of a square, according to one school, and a circle, according to a different school. The altar for the *Āhavanīya* should be always square and that of the *Dakṣiṇa* semi-circular. The area of each, however, must be the same and equal to one square *vyāma* (1 *vyāma* = 96 *aṅgulis*). So the construction of these three altars, it will easily be recognised, pre-supposes the knowledge of the following geometrical operations:—

(*i*) To construct a square on a given straight line.
(*ii*) To circle a square and *vice versâ*.
(*iii*) To double a circle.

The last problem is the same as to evaluate the surd $\sqrt{2}$. Or it may be considered as a case of doubling a square and then circling it. So in that case, we get at the proposition :—

(*iv*) The area of the square on the diagonal of a square is double the area of that square.

The *Saumikī-vedi* or *Mahā-vedi* is described as an isosceles trapezium whose face is 24 padas (or prakramas), base 30 padas and altitude 36 padas. The *Sautrāmaṇī-vedi* is stated to be an isosceles trapezium similar to and with an area one-third that of the *Mahā-vedi*, and the *Paitṛkī-vedi* is one-ninth of the latter. The *Prāgvaṁśa* is a rectangle. These and other similar altars lead to the operations :

(*v*) To construct a rectangle having given sides.

(*vi*) To construct an isosceles trapezium whose face, base and altitude are given.

(*vii*) To find the area of an isosceles trapezium.

(*viii*) To construct an isosceles trapezium whose area will be equal to a simple multiple or sub-multiple of, and which will be similar to, another isosceles trapezium.

Geometrical operations of more complex nature are required for the accurate construction of the *Kāmya Agni* (or "the fire altars for the sacrifices to achieve special objects"). Amongst them the most ancient and primitive form is the *Śyena-cit* (or "the altar of the form

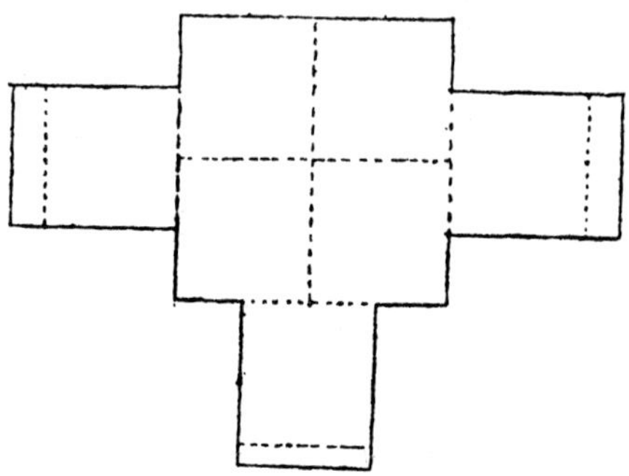

Fig. 1

of the falcon"). The *ātman* (or "body," "trunk") of this *citi* (or "altar") consists of four squares of one square puruṣa each. Each of its wings is a rectangle of one puruṣa by one puruṣa and one *aratni* (= $\frac{1}{5}$ of a puruṣa). Its tail is a rectangle of one puruṣa by one puruṣa and a *prādeśa* (= $\frac{1}{10}$ of a puruṣa). This altar is more usually called *Saptavidha-sāratni-prādeśa-caturasra-śyena-cit* because its area is $7\frac{1}{2}$ square puruṣas, its shape resembles that of a falcon (*śyena*) and because the bricks used in its construction are square.

The Fire-altars for other optional sacrifices are prescribed to be of different shapes. Thus we find altars also of the shape of (2) *vakra-pakṣa vyasta-puccha śyena* (or "the falcon with bent wings and outspread tail"), (3) *kaṅka* ("heron"), (4) *alaja* (a kind of bird), (5) *praüga* ("triangle" usually an isosceles triangle), (6) *ubhayataḥ praüga* ("triangles on both sides," that is, a rhombus), (7) *ratha-cakra* ("chariot wheel"), (8) *droṇa* ("trough"), (9) *samuhya* ("combined"), (10) *paricāyya* ("circular"), (11) *śmaśāna* ("cemetery"), (12) *kūrma* ("tortoise"), etc. Each of these altars shall have the same area as that of the standard form of the *Śyena-cit*, that is, $7\frac{1}{2}$ square puruṣas.

For the accurate construction of these altars, previous knowledge of the following principal geometrical propositions will be essential besides those noted above and a few others:

(*ix*) To construct a square equal to a simple multiple (or sub-multiple) of another square.

(*x*) To construct a square equal to the sum or difference of two unequal squares.

(*xi*) To transform a rectangle into a square and *vice versâ*.

(*xii*) To construct a triangle or a rhombus equal to a square.

A knowledge of the following important theorem is most indispensable for the geometry of the altar-construction.

(*xiii*) The area of the square described on the diagonal of a rectangle is equal to the sum of the areas of the squares described on its two sides.

Every one of the altars is constructed with five layers of bricks, which together come usually up to the height of the knee (=32 *aṅgulis*). In some cases the use of more layers of bricks is permitted with the proportional increase in the height of the altar. Now every layer, it is prescribed, contains a definite number of the bricks of specified shapes. For instance, each layer of the square *Gārhapatya* altar is constructed with 21 bricks of square or rectangular shape and each layer of the *Caturasra Śyena-cit* consists of 200 square bricks. Again in the case of the altars of other optional sacrifices, shape of the bricks are varied, but the number of them to be employed in the construction remains the same, *i.e.*, 200. Sometimes the one and the same altar is constructed in different patterns. All these have given rise to (1) the problems of the division of figures into a particular number of parts of specified shapes and also to (2) certain interesting problems of indeterminate character.

It has been stated above that a *Kāmya Agni* has an area of 7½ square puruṣas. That is the case only at the first construction of the altar. At its second construction, the area has to be increased by one square puruṣa; at the third construction by two square puruṣas; and so on until to the size of 101½ square puruṣas. But the strict injunction of the scriptures is that the shape of the altar on the whole, that is, the relative proportion between its different constituent parts at any construction, must not be altered. Thus arise the problems of constructing similar figures.

Such is, in brief, a résumé of the more salient points in the elaborate and minute in details specifications of the shape and size of the principal sacrificial altars and of the geometrical knowledge presupposed in their construction, as we find them in the extant *Śulba*. What should be particularly emphasized now is the fact that those specifications are not due to the authors of the *Śulba* themselves. They do not even pretend to make any such claim. On the other hand, they have often and then expressly admitted to have taken them from earlier works. We, in fact, find that numerous passages of *Baudhāyana* and *Āpastamba Śulba* dealing with the spatial magnitudes of sacrificial altars as well as with the methods of their construction, end with the remark *iti vijñāyate* [or "it is known," "it is recognised or prescribed (by authorities)"].[1] Sometimes *iti abhyupadiśanti* (" thus they teach ")[2] or *iti uktaṃ* ("it has been said"),[3] is used in the same sense. It has been rightly pointed out before by Garbe[4] that all those passages of Āpastamba are literal quotations from the *Taittirīya Brāhmaṇa* or from the Brāhmaṇa-like portions of the *Taittirīya Saṁhitā* or *Āraṇyaka*. That is exactly true also of the similar passages of Baudhāyana.[5] This writer is occasionally more explicit about his sources. In connexion with certain difference of opinions amongst the altar-builders about the proper size and shape of a

1 *BSl*, i. 65, 71, 76, etc. ; *ĀpSl*, iv. 1, 3, 5 ; v. 1, 8, 10, etc.

2 *BSl*, i. 85.

3 *ĀpSl*, ix. 2.

4 *Vide* the Preface (p. xviii) to his edition of the *Śrauta Sūtra of Āpastamba*, Vol. III, Calcutta, 1902. Garbe has pointed out in a most scholarly manner the relations of this work with others such as *Saṁhitā*, *Brāhmaṇa* and *Śrauta-sūtra*.

5 Compare for instance the passages with such remarks in *BŚr*, xxiv. 2 with *TS*, i. 2. 2. 3; *BŚr*, xxiv. 29 = *TS*, i. 7. 3. 1 ; *BŚr*, xxvi. 21 = *TS*, vii. 4. 2. 3, *PañBr*, xxiii. 19. 8 ; etc.

particular altar,[1] Baudhāyana is found appealing to the authorities of the *Brāhmaṇa*, by name, for the purpose of arriving at a satisfactory settlement. "This is not right," observes he, "as it will bring this opinion in contradiction with the ancient precepts. Regarding this point the *Brāhmaṇa* of some is as follows..., of others is...And the following is our *Brāhmaṇa*..."[2] By "our Brāhmaṇa" is meant the *Taittirīya Saṁhitā*, where indeed the quoted passage occurs.[3] On a different occasion, in connexion with certain method of constructing a particular altar, Baudhāyana remarks: "There is also a *Brāhmaṇa* on this point."[4] Here again the reference is to the *Taittirīya Saṁhitā*.[5] There are also other mentions of *Brāhmaṇa* in general by Baudhāyana.[6] He has once quoted the *Maitrāyaṇīya Brāhmaṇa* by name.[7] Kātyāyana is found to have appealed similarly to the authority of the "Śruti," on two occasions.[8] Āpastamba has sometimes observed that certain constructions are not sanctioned by the *Śruti*.[9] Thereby he clearly implies that other matters about the spatial magnitudes of the sacrificial altars and the methods of constructing them, that have been recorded by him, are in full accordance with the teachings of the *Śruti* but are not his devices. This he has admitted

1 The controversy is as regards the construction of the falcon-shaped altar of areas $1\frac{1}{2}$ to $6\frac{1}{2}$ square puruṣas with or without wings and tail. This will be dealt with more fully later on. Compare *ĀpŚl*, viii. 3-5.

2 *BŚl*, ii. 15-9.

3 *TS*, v. 2. 5. 1.

4 *BŚl*, ii. 35.

5 *TS*, v. 6. 6. 3.

6 *BŚl*, iii. 6. Here the reference is to *TS*, v. 3. 1. 5 and v. 5. 3. 2. Compare also *BŚl*, iii. 1 with *TS*, v. 4. 11. 1.

7 *BŚl*, iii. 10.

8 *KŚl*, v. 7; vi. 4.

9 *ĀpŚl*, viii. 5, 6.

also otherwise, as has been just pointed out. It will be further shown presently that we can, indeed, trace most of the matters contained in the *Śulba* to the earlier *Brāhmaṇa* and *Saṁhitā*.

The reference to the sacrificial altars and their construction is found as early as the *Ṛg-veda Saṁhitā* (before 3000 B.C.).[1] There is mention in that work of the "three places" of the Agni,[2] which doubtless imply the *Gārhapatya*,[3] *Āhavanīya* and *Dakṣiṇāgni*. Though we do not find there any specific mention about the relative sizes and shapes of these altars we have nothing to doubt that they were, in any way, different from what we meet with in posterior *Brāhmaṇa*.[4] Hence it seems that the problem of the squaring of the circle and the theorem of the square of the hypotenuse (at least in its simplest form) are as old in India as the time of the *Ṛg-veda*. They might be older still. For it has been shown by Oldenberg that those three fires are earlier than the *Ṛg-veda*.[5]

In the *Ṛg-veda*, it should be made clear, there is no particular rule for the construction of the altars. We cannot indeed reasonably expect to find such a rule there,

[1] There are innumerable references in the *Ṛg-veda* to the sacrifice-altars and their constructions. For the mention of the *vedi*, see for instance *ṚV*, i. 164. 35; i. 170. 4; v. 31. 12; vii. 35. 7, etc.; and for its construction compare the passages: "O Lovely (*Agni*)! They construct the *vedi* for you and offer oblations there" (*ṚV*, viii. 19. 18); "measured out the *vedi*" (*ṚV*, x. 61. 2), etc.

[2] "यज्ञस्य केतुं प्रथमं पुरोहितमग्निं त्रिषधस्थे समीधिरे," *ṚV*, v. 11. 2.

[3] The mention of the *Gārhapatya* Fire by name occurs, for instance, in *ṚV*, i. 15. 12; vi. 15. 19 and x. 85. 27.

[4] The first express description of the *Gārhapatya* as a circle of one square *vyāma* (= *puruṣa*) and of the *Āhavanīya* being a square of the same size appears in the *Śatapatha Brāhmaṇa* (vii. 1. 1. 37; vii. 2. 2. 1 ff. *Cf*. *SBE*, Vol. XL, iii, p. 307, fn. 2). In the *Taittirīya Saṁhitā* (v. 2.5.1), the *Āhavanīya* is stated to be of one (square) puruṣa.

[5] Oldenberg, *Religion des Veda*, p. 348, n. 2; *SBE*, Vol. XXX, p. ix.

when we remember the nature of that work. We learn from it, however, that there were then learned experts to do that work. This is very important inasmuch as it implies the existence of the science or the art of the altar-construction. An expert in that science is called *Agnicit* (or "the constructor of the *Agni* or Fire-Altar"). This term appears in the *Taittirīya Saṁhitā*, *Maitrāyaṇīya Saṁhitā*, *Śatapatha Brāhmaṇa* and other early works In these works we also find rules for his conduct.[1]

We next turn to the *Yajur-veda* which is very rightly described as the encyclopædia proper of the Vedic sacrifices. There we find a very elaborate, and in detail tedious, rite of the *Agni-cayana* (or "the construction of the Fire-altar") and its highly speculative philosophy.[2] The same mystic import is found, it is noteworthy, in the *Saṁhitā* of the different schools of this Veda such as the *Taittirīya*, *Maitrāyaṇīya*, *Kāṭhaka*, *Kapiṣṭhala* and *Vājasaneya*. This shows that the Agni-cayana and its philosophy had taken definite shapes in an earlier period. Eggeling[3] has, indeed,

[1] According to the *Taittirīya Saṁhitā*, an Agnicit should live upon what are obtained freely from Nature, such as fruits, etc., but not by sowing (v. 2. 5. 5-6). He is particularly forbidden to eat the flesh of birds. "The fire is a bird; if the piler of the fire were to eat of a bird, he would be eating the fire, he would go to ruin" (*TS*, v. 7. 6. 1). The same prohibition is found also in the *Maitrāyaṇīya Saṁhitā* (iii. 4. 8) and *Śatapatha Brāhmaṇa*. In the latter work we find also a contrary opinion. "Here, now, they say, 'He who has built an altar must not eat of any bird, for he who builds a fire-altar becomes of a bird's form; he would be apt to incur sickness: the Agnicit therefore must not eat of any bird.' Nevertheless, one who knows this may safely eat thereof; for he who builds an altar becomes of Agni's form, and, indeed, all food here belongs to Agni: whosoever knows this will know that all food belongs to him." (*ŚBr*, x. 1. 4. 13. Compare also *TS*, v. 6 .8. 3 f.; 7. 6. 2 ff.; *ĀpŚr*, xvii. 24.

[2] Compare *The Veda of the Black Yajus School* translated into English by A. B. Keith, Cambridge, Mass., 1914, Introduction, p. cxxv ff.

[3] *SBE*, Vol. XLIII, Introduction, pp. xiv ff.

traced the origin of the philosophy to the *Ṛg-veda.* In the *Brāhmaṇa*, the science of the construction of the Fire-altar is found in an enormously developed form. Thus five sections (*kaṇḍikā* vi-x) out of a total of fourteen, or rather more than one-third of the whole of the *Śatapatha Brāhmaṇa* is devoted to the treatment of this science. We cannot definitely ascertain how much of this development is due to the *Brāhmaṇa* and how much of it is still older. For, according to the introductory chapters of the *Hiraṇyakeśi* and *Āpastamba Śrauta-sūtra*, it is one of the primary functions of the *Brāhmaṇa* to describe, amongst other things, the *Karma-vidhi* (or "Rules for the performance of the sacrificial rites") and the *Purā-kalpa* (or "the performance of the sacrifices in former times"). So that the *Brāhmaṇa* simply keeps up a record of the ancient traditions. Indeed in those works, all rites are traced to the gods as their originators or even to the Supreme Creator of the universe. Still the general truth is that every science has its growth and development and the science of the altar-construction cannot be an exception to it. So it will be quite natural to believe—and we have corroborative evidence of it—that the science truly attained a new stage in the time of the *Brāhmaṇa* (c. 2000 B.C.). The existence of different masters of this science with independent views is found even as early as the time of the *Taittirīya Saṁhitā* (c. 3000 B.C.).[1] More conclusive evidence of it is furnished by the *Śrauta-sūtras* which are a sort of compendiums of earlier theories. In the *Śatapatha Brāhmaṇa*, there are clear attempts to refute some of the earlier theories of the science of the altar-construction (*vide infra*).

[1] References to the earlier authorities are found, for instance, in *TS*, v. 2. 8. 1-2; 3. 8. 1; 5. 2. 1 ff.; etc.

One thing should be made here perfectly clear: the treatment in the earlier literatures, the *Saṁhitā* as well as the *Brāhmaṇa*, of the measurement of the various *Vedi* and *Agni*, appertains chiefly to the ritualistic aspects of the problems. Reference to the secular or geometrical and other truly scientific aspects are only incidental for them and hence are found on rare occasions. Fuller details of the geometry of the measurements of the altars are particularly described in the *Śulba* parts of the *Śrauta-sūtra*. But traces of that, it will be shown conclusively in the course of this work, are clearly noticeable also in the *Brāhmaṇa*. And it will not be improper, I think, to presume that the geometrical methods for the solution of the problems of the measurements of the altars were known in still earlier periods. For the rituals of measurements will be altogether baseless unless accompanied by a knowledge of the underlying geometry.

In the *Taittirīya Saṁhitā*, we find the following scanty reference to the scientific operations for the construction of the *Dārśapaurṇamāsikī-vedi*:

"He performs the second drawing of a boundary himself. The earth is of the size of the altar; verily having excluded his enemy from so much of it, he performs the second drawing of a boundary himself. Cruelly he acts in making an altar."[1]

But such meagre descriptions of course do not help us in any way to conjecture the geometrical devices adopted for the construction.

We shall now proceed to show, as briefly as possible, that some of the specifications about the shape and size of the various *Vedi* and *Agni* and about their relative positions, which we find in the *Śulba* can be clearly traced

[1] *TS*, ii. 6. 4. 2-3 (Keith's translation). Compare *TBr*, iii. 2. 9. 10; *BŚr*, xxiv. 24.

to earlier *Saṁhitā* and *Brāhmaṇa*. This will doubtless corroborate the traditional origin of the science of Hindu Geometry in a very remote age to be true. We have already shown the ancient origin of the three fundamental altars, the *Gārhapatya*, *Āhavanīya* and *Dakṣiṇāgni*. Their relative positions are described in the *Śatapatha Brāhmaṇa*[1] and *Śrauta-sūtra*[2] to be identical. According to all the *Saṁhitā*[3] and *Brāhmaṇa*,[4] as in the *Śulba*, the *Gārhapatya citi* must be constructed with the same number of bricks, namely 21, arranged in an identical manner. It is further stated in the *Taittirīya Saṁhitā*:

"He who constructs (the *Gārhapatya citi*) for the first time should construct in five layers ... He who constructs for a third time should construct in one layer...."[5]

The spatial magnitudes of the *Saumikī-vedi* (or "the altar of the Soma-sacrifice"), also called the *Mahā-vedi* ("the Great Altar") which has been already described to be of the form of an isosceles trapezium whose face is 24 *prakramas* (or *padas*) long, base is 30 and altitude 36 *prakramas*, are given in the *Saṁhitā*,[6] and *Śatapatha Brāhmaṇa*.[7] But the earliest description of a method of its measurement, or a method for the construction of an isosceles trapezium having given face, base and altitude is found in the latter work. It says:

"From that (the largest post on the east side) one proceeds three vikramas to the east and there fixes a pole; this

---

1 *ŚBr*, i. 7. 3. 23-5.

2 *Cf.* *BŚr*; *BŚl*, i. 64-69; *ĀpŚr*, v. 4. 3-5; *ĀpŚl*, iv. 1-4; *KŚr*, iv. 8. 19; *KŚl*, i. 26,2 8.

3 *Cf.* *TS*, v. 2. 3. 4ff; *MS*, iii. 2. 3; *KṭS*, xx. 1; *KapS*, xxxii. 3.

4 *ŚBr*, vii. 1. 1. 18, 33-4.

5 *TS*, v. 2. 3. 6f.

6 *TS*, vi. 2. 4. 5; *MS*, iii. 8. 4; *KṭS*, xxv. 3; *KapS*, xxxviii. 6.

7 *ŚBr*, iii. 5. 1. 1ff; x. 2. 3. 4. Compare also *BŚr*, vi. 22; *ĀpŚr*, xi. 4. 11-6; *KŚr*, viii. 3, 6-12; *MāŚr*, ii. 2. 1-12.

is the middle-hind pole (*antaḥpātaḥ*). From the middle-hind pole, he goes 15 prakramas towards the south and fixes a pole there; this is the south-west corner (of the *Mahā-vedi*). From the middle-hind pole, he proceeds 15 prakramas towards the north and fixes a pole there; it is the north-west corner. From the middle-hind pole he goes 36 prakramas towards the east and fixes a pole; this is the middle-front pole. From the middle front pole, he strides 12 prakramas towards the south and fixes a pole; this is the south-east corner. From the middle-front pole, he goes 12 prakramas towards the north and fixes a pole there; this is the north-east corner. Such is the measurement of the (*Mahā-*)*vedi.*" [1]

Here or anywhere else in this *Brāhmaṇa*, we are not taught how to draw the east-west line and how to draw a line at right angles to it (*i.e.*, the north-south line) through a given point on it. That there were some methods for those constructions is beyond question. We have only to conjecture what were these methods. Now almost identical descriptions about the measurement of the *Mahā-vedi* reappear in the *Śrauta-sūtra* of Baudhāyana[2] and Āpastamba.[3] It has been further taught by the former writer that all the measurements are to be made by means of a cord on the principle of a rational rectangle (*vide infra*). The *Śatapatha Brāhmaṇa* is known to have measured the *vedi* with a cord.[4] Had it also recourse to the same method? At any rate, it is not improbable. For, we have clear evidence to prove

[1] *ŚBr*, iii. 5. 1. 1-6.

[2] *BŚr*, vi. 22. In this work the middle-hind pole is called the *śālā-mukhīya-śaṅku* and the middle-front pole the *yūpāvaṭīya-śaṅku.*

[3] *ĀpŚr*, xi. 4. 12-3. Compare also *KŚr*, viii. 3. 6-12, and the method of *KtŚr* quoted in Yājñikadeva's commentary on the 11th Sūtra.

[4] Cf. *ŚBr*, x. 2. 3. 8 ff.

that the theorem of the square of the diagonal, or the so-called Pythagorean Theorem was known then and used to be employed in that as well as in other connexions.[1]

The construction of a square having a given side is described thus (omitting the descriptions of the ceremonies and speculative explanations):

"He then takes up the wooden pin (*śamyā*) and wooden sword (*sphya*). Then from the pole which lies in the north-east (corner of the *Mahā-vedi*) strides three prakramas backwards and then marks out the pit (*cātvāla*). That is the measure of that pit; it has no other measure. Wherever he himself thinks it (proper) in his mind, in front of the *utkara* ('the heap of rubbish'), there he marks out the pit. He (draws first) the (western) extremity of the altar. He lays out the wooden pin northwards and marks out (a line)...Then on the front: he lays down the wooden pin northwards and marks out (a line)...Then on the southern extremity of the altar: he lays down the wooden pin eastwards and marks out (a line)...Then on the north: he lays down the wooden pin eastwards and marks out (a line)...."[2]

This is, in fact, the square pit, with the earth from which the *Uttara-vedi* is constructed. Hence both have the same cubical content. This measurement of the pit reappears in the *Sulba*. It is, perhaps, particularly noteworthy, that in the above we find an instance of the use, in former times, of a ruler (in the body of the straight wooden pin, called the *śamyā*) to draw a straight line from a given point in a specified direction. It is not said how those directions, or rather the cardinal directions passing through a point were used

[1] *Vide infra.*

[2] *ŚBr*, iii. 5. 1. 26-30. Compare also *TS*, vi. 2. 7. 1-2.

to be determined. Was a pair of compasses also in use then?

Though lavish description of the rites and ceremonies in connexion with the construction of the various other altars, such as the *Darśapaurṇamāsa-vedi*, *Uttara-vedi*, *Aśvamedha-vedi*, *Agnīdhrīya*, *Hotrīya*, *Mārjālīya*, *Sadas*, *Uparavas*, etc., are commonly found in the *Taittirīya* and other *Saṁhitā*, any clear mention of their spatial magnitudes are very rare therein. In that respect, we obtain much better information from the *Brāhmaṇa*.

It has been stated before that the standard form of an optional Fire-altar is that of a certain bird. This bird is called *Śyena* ("falcon") in the *Taittirīya Saṁhitā*[1] and *Suparṇa Garutman* ("well-winged eagle") in the *Vājasaneya Saṁhitā*[2] and *Śatapatha Brāhmaṇa*[3] which is sometimes abridged into *Suparṇa* in the latter.[4] The first name is found more commonly in other *Saṁhitā* and *Śrauta-sūtra* whereas the other names are rarely met with elsewhere.[5] A clear reference to this form is found in the *Ṛgveda* where Agni is frequently called a bird.[6] The spatial magnitudes of the falcon-shaped Fire-altar have been defined in almost all the earlier works from the *Taittirīya Saṁhitā* onwards, and they are exactly the same as those that are found in the *Śulba*. Though it is stated by the authorities that it should be measured preferably with a bamboo-rod the details of the method of

1 *TS*, v. 4. 11. 1.

2 *VS*, xii. 4.

3 *ŚBr*, x. 2. 2. 4.

4 *ŚBr*, vi. 7. 2. 6. 8.

5 The name *Suparṇa-citi* of the *Śatapatha Brāhmaṇa* (vi. 7. 2. 8) reappears in the *Mānava Śulba* (vii), but its form differs from the ancient one by the addition of a head.

6 *RV*, i. 164. 52; x. 14. 5; compare also i. 58. 5; 141. 7; ii. 2.4; vi. 3. 7; 4. 7; x. 8. 3.

measurement are scanty in earlier ones. The *Taittirīya Saṁhitā* says:

"With man's measure he metes out; man is commensurate with the sacrifice; verily he metes him with a member of the sacrifice; so great is he as a man with arms extended; so much strength is there in man; verily with strength he metes him. Winged is he, for wingless he could not fly; these wings are longer by an ell (*aratni*); therefore birds have strength by their wings. The wings and the tail are a fathom (*vyāma*) in breadth; so much is the strength in man, he is commensurate in strength. He metes with a bamboo;..." [1]

The *Maitrāyaṇīya Saṁhitā* says:

"As much as a man with arms extended, with so much a bamboo-rod, (the Fire-altar) is meted out; so much strength is there in the man; verily with strength it is meted out ;...He metes out the Fire-altar; seven (square) puruṣas he metes out; for by seven puruṣas he knows the universe and by seven puruṣas of the self he eats food. A measure of *aratni* is added to the two wings; the birds have strength by their wings." [2]

More particulars are supplied by the *Śatapatha Brāhmaṇa*:

"Verily He comprises seven puruṣas. Seven puruṣas certainly are in this Person (Agni); since four (puruṣas) (as) the body and three the wings and tail; for the body of that Person is certainly (composed of) four (puruṣas) and the wings and tail of three. He metes it out with (the measure of) a man (*puruṣa*) with arms extended. Verily the sacrifice is a puruṣa and hence by it, all these are measured ; and that is its best

[1] *TS*, v. 2. 5. 1ff. (translation by Keith).

[2] *MaiS*, iii. 2. 4.

measure inasmuch as with arms extended he (man) has his maximum measure: he then secures for him that and by that he measures it... Then he adds two aratnis to the two wings; by that he gives strength to the wings. Verily, the two wings are two arms (of the bird) and by arms food is eaten: simply for the sake of food he makes that space; inasmuch as the food is taken from the distance of an aratni the two aratnis he adds to the two wings. Then to the tail he adds a vitasti. He thus gives strength to the support; verily, the tail is the support. The hand (consists of) vitastis, and by means of the hand the food is eaten; simply for the sake of food he makes that space. Inasmuch as he adds one vitasti to the tail, he settles for it the food; because he adds less here (in the tail), he thereby secures it in the food. Thus, this much is it (the body) measured, and this much is it (wings and tail); certainly it (the bird or altar) is measured this much in order to secure for it that (its natural measure)."[1]

Full details of the methods of measurement of the falcon-shaped Fire-altar are not found until the *Śrauta-sūtra*. The method of the measurement by means of a bamboo-rod has been described in the *Āpastamba Śulba*,[2] and that by means of a cord is hinted in the *Baudhāyana Śrauta*[3] described in full in the *Kātyāyana Śrauta*. The latter says:

" Measure a cord two puruṣas long. Make ties at its both ends. Make marks at the middle; on either sides of it, at the halves of the puruṣas; at distance of the one-fifth of a puruṣa from the middle (mark); and also

1 *ŚBr*, x. 2. 2. 5-8. *Cf.* vi. 1. 1. 6. ; x. 2. 3. 4.

2 *ĀpŚl*, viii. 7-ix. 3. Compare also *ĀpŚr*, xvi. 17.8.

3 *BŚr*, x. 19; xix. 1.

at their halves (that is, at one-tenth puruṣa from the middle mark). Stretch the cord along the *pṛṣṭhyā* (the east-west line) and fix poles at the two ties, the middle mark and the marks at semi-puruṣas. Unfasten the two ties, fasten them to the semi-puruṣa-poles and then stretch the cord towards the south by holding it by the middle mark. Make a point at the place reached by that. Unfasten the two ties; fasten one at the middle pole, then stretch the cord towards the south over the point and fix a pole at the place reached by the middle mark. Then fasten one tie at this pole and another at the eastern pole; stretch the cord towards the south and fix a pole at the place of the middle mark; then another also at the semi-puruṣa-mark. Unfasten the tie from the eastern pole and then fasten it to the western pole; stretch the cord southwards and fix a pole at the place reached by the middle mark and also two about it at the semi-puruṣa marks. Proceed in the same way on the northern side. Again on the southern side, stretching the cord in the way indicated before, fix a pole at the distance of the fifth-puruṣa-mark. Having fastened a tie at that, and also at the eastern semi-puruṣa pole, stretch the cord properly, and fix a pole at a distance of the semi-puruṣa-mark (from the one) and the fifth-puruṣa-mark (from the other). Similarly on the west. Similarly (construct) the northern wing. Thus also the tail with its vitasti. If desired, the two sides of each wing and of the tail may be contracted by four aṅgulis each on one extremity and extended by the same amount on the other."[1]

The *Śatapatha Brāhmaṇa*, however, teaches us how to bend the wings of the falcon in order to construct that

[1] *KŚr*, xvi. 8. 1-20.

variety of the Fire-altar, known as, the *Vakrapakṣa Śyena-citi.*

" He contracts the inner extremity (of the southern wing) inside on both sides only by four aṅgulis; by four aṅgulis outwards on both sides he expands the outer extremity. Thus by as much he contracts, by so much he expands; certainly, for that, he neither exceeds (the proper size of the wing), nor makes it too small. Similarly, he does for the tail; and in the same way, for the northern wing. Then he makes the bent of the two wings. For bents there are in the wings of a bird; the bents of the wings of a bird are by its one-third each; by one-third of the wings inwards each the bents of the wings of the bird are. He expands (each of the wings) on the front just by four aṅgulis; he contracts at the back by four aṅgulis. Thus by as much he expands by so much he contracts; and so he neither exceeds, nor makes it too small."[1]

A complete list of the various *Kāmya Agni* together with a statement of the objects for the attainment of which, each of them is to be constructed and sacrifices made therein, is found in the *Taittirīya Saṁhitā.*[2] That has been practically reproduced in the *Baudhāyana Śrauta-sūtra.*[3] The enumeration of most of them appears in the *Maitrāyaṇīya Saṁhitā*[4] and *Śatapatha Brāhmaṇa.*[5] It may be noted, though it is immaterial for our purpose, that the construction of Fire-altars other than the *Suparṇa-citi* (" the eagle-shaped altar ") is forbidden in the latter work.

[1] *ŚBr*, x. 2. 1. 4-5; compare also *ŚBr*, x. 2. 1. 7; *BŚl*, iii. 62ff.

[2] *TS*, v. 4. 11.

[3] *BŚr*, xvii. 28-30.

[4] *MaiS*, vii. 4. 7. *Cf.* iii. 2. 5.

[5] *ŚBr*, vi. 7. 2. 8.

Thus we find that almost all the *Vedi* and *Agni* which are described in the *Śulba* can be traced back, for the matter of their shapes and sizes, as far as the time of the *Brāhmaṇa* (c. 2000 B.C.); and they are mentioned even in the *Saṁhitā* (c. 3000 B.C.). The *Śulba* has, in fact, expressly admitted in the majority of cases, as has been pointed out before, that it has taken the spatial magnitudes of the altars from the earlier literatures. We can, similarly, trace the earlier origin of many other matters traced in the *Śulba*. As regards the height of an Agni and the number of bricks to be used in its construction, *Taittirīya Saṁhitā* observes:

"He should pile (the fire) of a thousand (bricks) when first piling (it); this world is commensurate with a thousand; verily, he conquers this world. He should pile (it) of two thousand when piling a second time; the atmosphere is commensurate with two thousand; verily, he conquers the atmosphere. He should pile (it) of three thousand when piling for the third time; yonder world is commensurate with three thousand; verily, he conquers yonder world. Knee-deep should he pile (it), when piling for the first time; verily, with the Gāyatrī he mounts this world; navel-deep should he pile (it) when piling for the second time; verily with the Triṣṭubh he mounts the atmosphere; neck-deep should he pile (it) when piling for the third time; verily, with the Jagatī he mounts yonder world."[1]

Each Agni is usually constructed in five layers, when constructed for the first time. It should have double or treble number of layers when constructed for the second or third time. The *Taittirīya Saṁhitā* says:—

[1] *TS*, v. 6. 8. 2 f. (Keith's translation).

" The first layer is this (earth), the mortar the plants and trees; the second is the atmosphere, the mortar the birds; the third is yonder (sky), the mortar the Nakṣatras; the fourth the sacrifice, the mortar the sacrificial fee; the fifth the sacrificer, the mortar the offspring; if he were to pile it with three layers, he would obstruct the sacrifice, the fee, the self, the offspring; therefore should it be piled with five layers; verily, he preserves all. In that there are three layers, (it is) since Agni is of threefold; in that there are two (more), the sacrificer has two feet, (it is) for support; there are five layers, man is five-fold; verily, he preserves himself. There are five layers, he covers (them) with five (sets of) mortar, these make up ten, man has ten elements; he preserves man in his full extent."[1]

But when *Nākasad* and *Pañcaşoḍā* bricks are employed, after the fourth layer, there will be a sixth layer as the height of these bricks are half the usual height of a brick, *viz.*, the one-fifth of a *jānu* (=32 aṅgulis). This sixth layer is mentioned also in the *Taittirīya Saṁhitā*.[2]

The growth and development of (1) the theorem of the square of the diagonal, (2) the quadrature of a circle, and (3) the construction of similar figures, has been treated elsewhere in their proper places. We have clear proofs, it has been shown there, of the use of these in the time of the *Śatapatha Brāhmaṇa* (c. 2000 B.C.). The first two seem to be still older. But we do not find an enunciation of the theorem of the square of the diagonal and a method for the quadrature of the circle before the time of the *Śrauta-sūtra*. A method for the construction of similar figures is taught in the *Śatapatha Brāhmaṇa*, and it is the same as we find in later works.

[1] *Ibid*, v. 6. 10. 2 f. (Keith). Compare *BŚl*, ii. 13.

[2] *Ibid*, v. 6. 10. 3. Compare *BŚl*, ii. 28, 59.

CHAPTER IV

POSTULATES

For the geometrical operations described in the *Sulba*, the authors, we find, have tacitly assumed the truth of certain other results without any attempt to describe them beforehand or to indicate how they could be effected. These results we have called here postulates of the *Śulba*. They might not be postulates in the Euclidean sense of the term; but they can certainly be so called in accordance with the meaning given by Aristotle, namely "whatever is assumed, though it is a matter for proof, and used without being proved." Most of the postulates of the *Śulba* are concerning the division of figures, such as straight lines, rectangles, circles and triangles. A few of them are about other matters of importance.

(*a*) *A given finite straight line can be divided into any number of equal parts.*

In the geometry of the *Śulba*, it is oftentimes required to divide a given finite straight line into a specified number of equal parts. For example in one instance, the diameter of a given circle is divided into 8 equal parts, each of these again into 29 parts and so on into other number of divisions.[1] There are indeed numerous such instances.[2] Now it will be naturally asked how it was used to be done. Certainly not arithmetically. At least it is not always possible to do so. For we find instances of division of straight lines which cannot be expressed in terms of commensurate numbers. In circling a square, such a straight line has

[1] *BŚl*, i. 59.

[2] *BŚl*, i. 60, 68-9, etc.

to be divided into three parts[1] and in another case into twelve parts.[2] Sometimes the given straight line is such that the parts when expressed arithmetically will contain big fractions. Thus the side of a square of 96 *aṅgulis* has to be divided into 7 equal parts.[3]

(*b*) *A circle can be divided into any number of parts by drawing diameters.*

In the *Śulba*, we have several instances of the division of a circle into a specified number of parts. For instance, it is said that the *Dhiṣṇiya* may be square or circular in shape and one of them, *viz.*, the *Agnīdhrīya*, has to be divided into nine parts. Now in the case of the square shape of the altar, it is easily divided into 9 smaller squares by drawing cross-lines through the points of trisection of the sides. When it is circular, there is described a small circle about its centre and the annulus

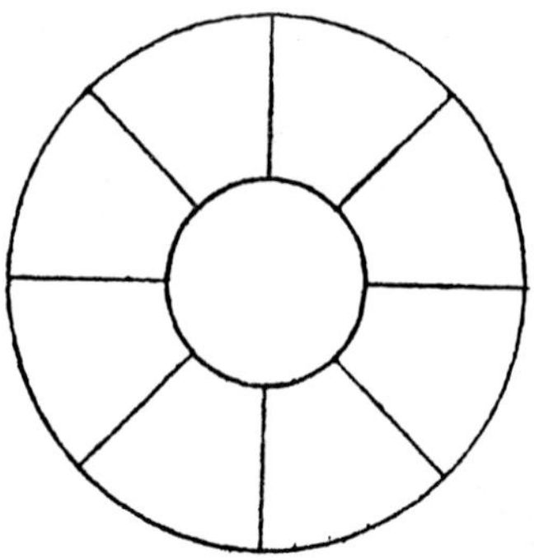

Fig. 2

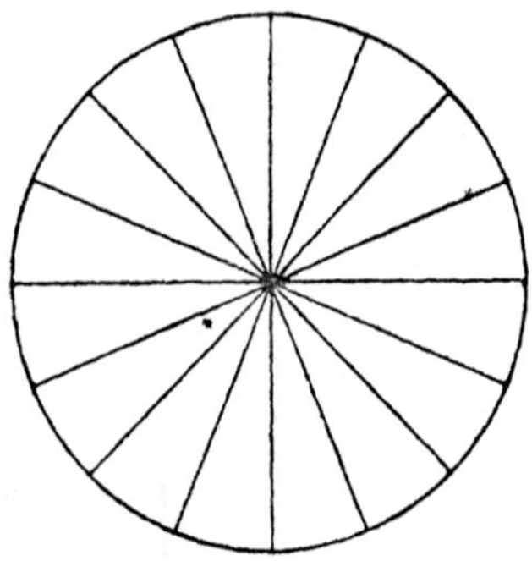

Fig. 3.

is then divided into 8 equal parts by drawing four diameters.[4] Similarly in an alternative circular shape of

[1] *BŚl*, i. 56.

[2] *BŚl*, iii. 162; compare also *ĀpŚl*, xix. 7 in which the diagonal of a square of sides $=\frac{1}{4}$ of a puruṣa is divided into seven equal parts.

[3] *BŚl*, ii. 64.

[4] *BŚl*, ii. 73-4; *ĀpŚl*, vii. 13-14.

the *Mārjālīya* fire, the circle has to be divided into 6 equal parts.[1] A circular annulus has been divided into 32 equal parts;[2] another annulus into 64 equal parts and then again into two parts each by drawing the mean circle.[3]

(c) *Each diagonal of a rectangle bisects it.*

(d) *The diagonals of a rectangle bisect one another and they divide the rectangle into four parts two and two vertically opposite of which are equal in all respects.*

The description of the division of a rectangle or a square by diagonals is found in the *Śulba* primarily in

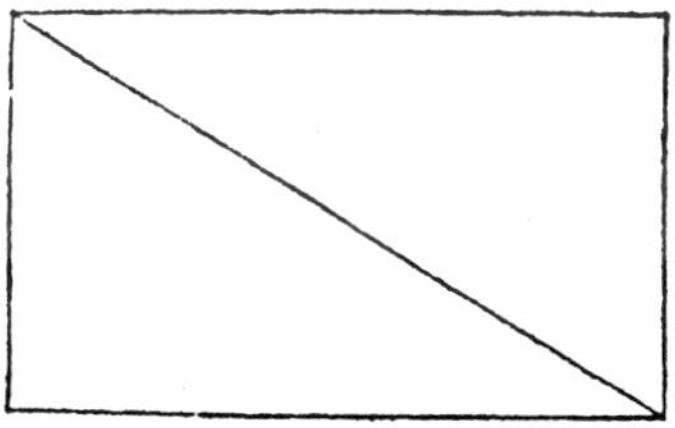

Fig. 4.

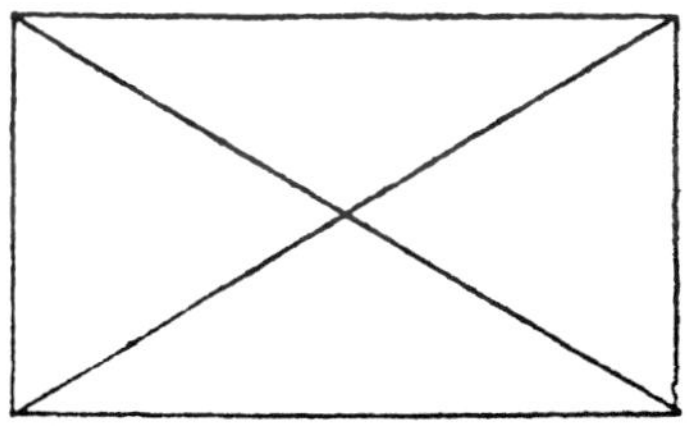

Fig. 5

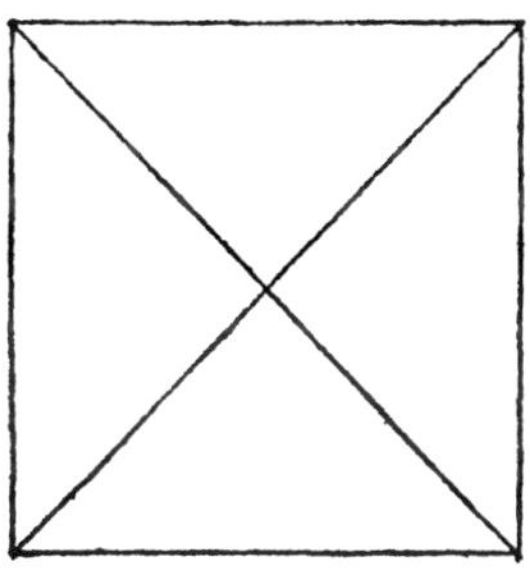

Fig. 6.

[1] *BŚl*, ii. 77. Compare *ĀpŚl*, xvii. 21. 3, 4 for instances of division of a circle into 12, 16 and 24 parts.

[2] *BŚl*, iii. 200.

[3] *BŚl*, iii. 202.

connexion with the making of bricks of desired size and shape. Certain interesting geometrical theorems have been assumed there. The brick which resembles in shape the portion of a rectangle or of a square divided by its diagonal is termed *ardhyā* (or "the half") and that resembling a portion by the two diagonals is called *pādyā* (or "the quarter"). There are distinguished two kinds of *pādyā* of a rectangle, *viz.*, *dīrgha-pādyā* (or "longish or broader quarter") and *śūla-pādyā* (or "trident quarter").[1] This distinctive nomenclature implies: (1) the halves of a rectangle or of a square by a diagonal are identical in size as well as in shape, and (2) so are also the quarters of a square by its diagonals; and (3) the diagonals of a rectangle divide it into four parts which are equal in area but they are of two kinds as regards their shape. These names perhaps further imply an idea of *obtuse* and *acute* angles. There are bricks which are halves of the quarter bricks by the perpendicular from the vertex on the base. No distinction is found to have been made between the half of a *dīrgha-pādyā* and that of a *śūla-pādyā*, which clearly shows that the Śulba-kāras were aware that those halves were identical. Thus it appears that the early Hindu geometers knew the simple cases of the congruence theorems.

Another interesting kind of bricks is formed by the combination of a half of a *dīrgha-pādyā* or a *śūla-pādyā* with another brick. Baudhāyana describes:

"The eighth parts of them[2] should be so combined as there will be (a brick having) three corners."[3]

[1] *BŚl*, iii. 168-9, 178.

[2] The reference is to a square brick, called *pañcamī*, for each side of it is equal to the fifth (*pañcama*) part of a puruṣa and to a rectangular brick one-fifth of a puruṣa by one-fifth of a puruṣa and its half.

[3] *BŚl*, iii. 122.

A brick of this kind is technically called *ubhayī*[1] (from *ubhay*, "both") because it is formed by the combination of

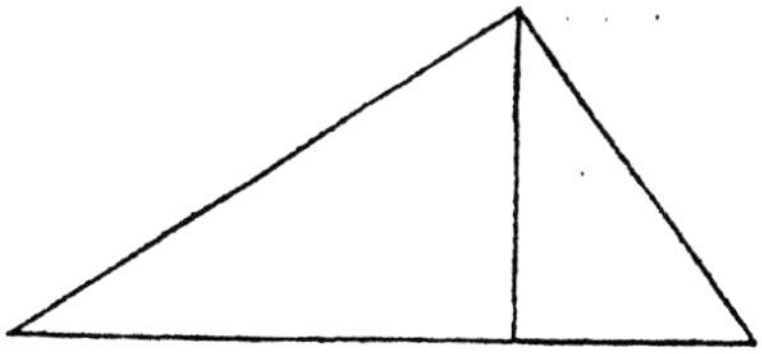

Fig 7

two bricks of two different kinds. Since there is mentioned only one *ubhayī* though there are distinguished two different kinds of quarter bricks of a rectangle (the *adhyardha*) it follows that Baudhāyana was fully aware, what has been just mentioned, that all the eighth parts of a rectangle are identical. What is much more noteworthy is that in the formation of the *ubhayī* we find the source of the discovery of the later Hindu principle of forming a rational *scalene* tringle by the juxtaposition of two rational right-angled triangles.[2]

(e) *The diagonals of a rhombus bisect each other at right angles.*

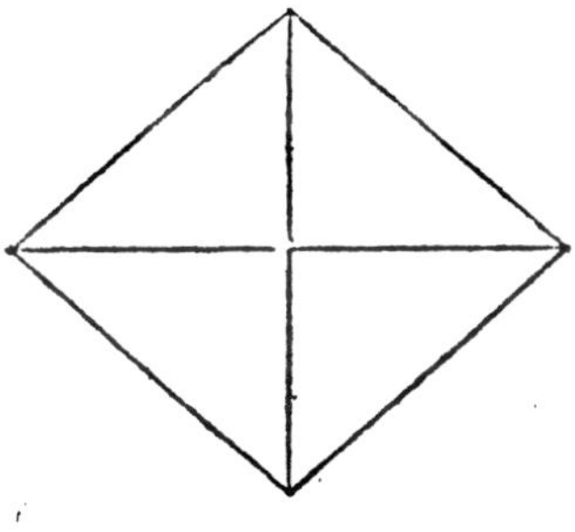

Fig.8

[1] *Ibid*, iii. 129.

[2] Bibhutibhusan Datta, "On Mahāvīra's Solution of Rational

(f) A triangle can be divided into a number of equal and similar parts by dividing the sides into an equal number of parts and then joining the points of division two and two.

Baudhāyana on a certain occasion says, "this (triangle) is divided into ten parts."[1] But how to do it he does not explain expressly. We, however, learn it from the commentators that the traditional practice in such a case was

Fig. 9.

to divide each side into four equal parts and then to join the points of division two and two as indicated in the Fig. 9.

(g) An isosceles triangle is divided into two equal halves by the line joining the vertex with the middle point of the opposite side.[2] Each of these has again been divided into six parts.[3]

Triangles and Quadrilaterals," *Bull. Cal. Math. Soc.*, Vol. xx, pp. 267-294; see particularly pp. 276-7.

[1] *BŚl*, iii. 256.

[2] *Ibid*, iii. 258.

[3] *Ibid*, iii. 260.

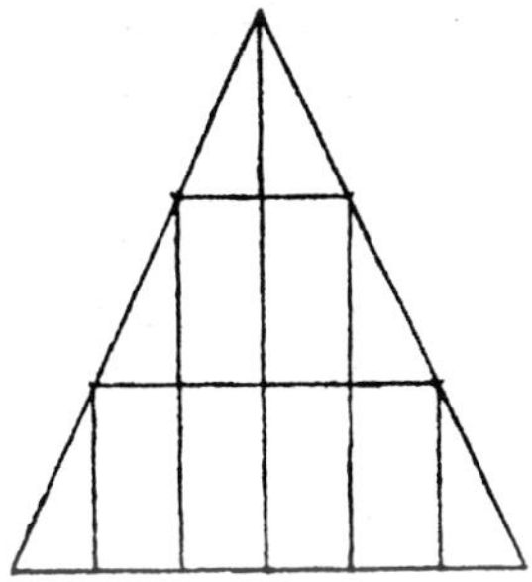

Fig 10

We shall see later on that figures of more complex shapes had to be divided into a specified number of parts, namely 200, of given forms. And this led to some interesting problems of indeterminate character.

(h) *A triangle formed by joining the extremities of any side of a square to the middle point of the opposite side is equal to half the square.*

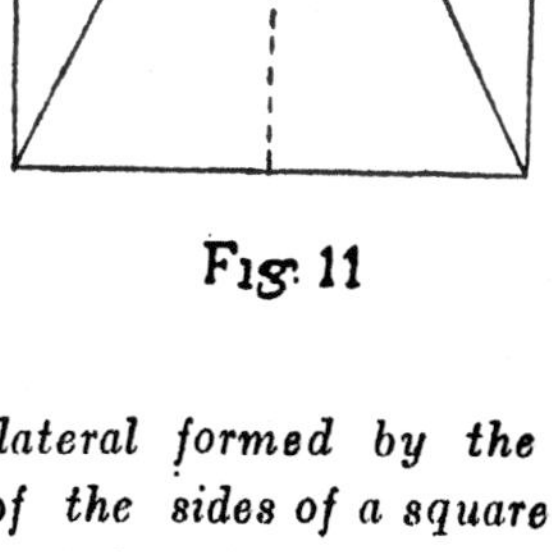

Fig 11

(i) *A quadrilateral formed by the lines joining the middle points of the sides of a square is a square whose area is half that of the original one.*

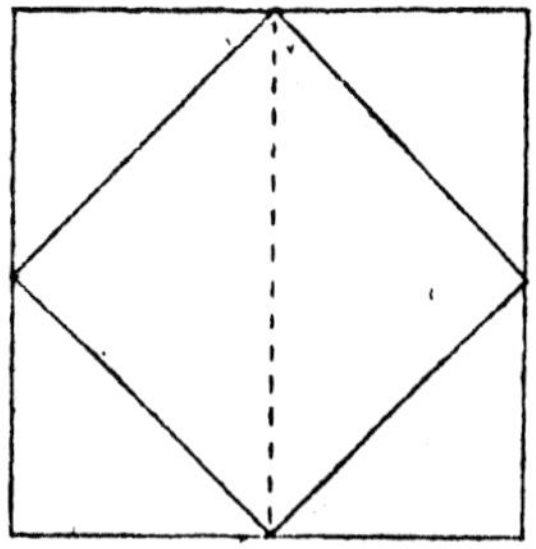

Fig. 12

(j) *A quadrilateral formed by the lines joining the middle points of the sides of a rectangle is a rhombus whose area is half that of the rectangle.*

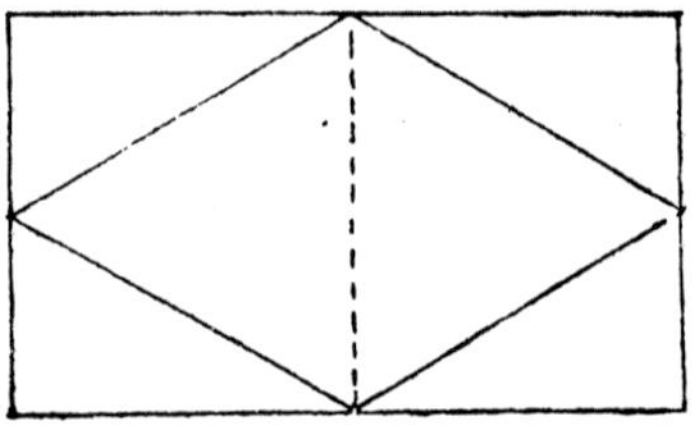

Fig. 13.

(k) *A parallelogram and a rectangle which are on the same base and within the same parallels are equal to one another.*

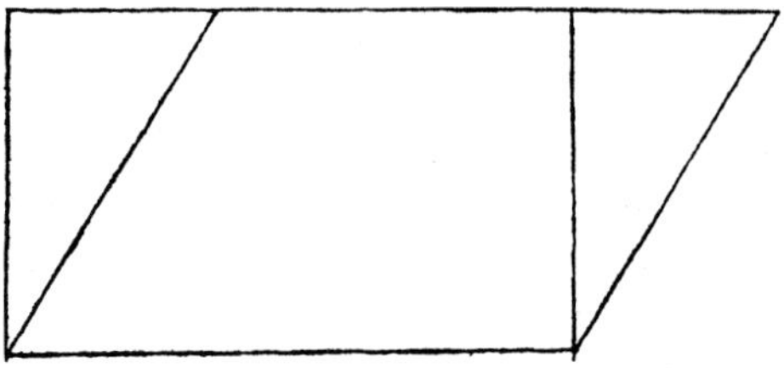

Fig. 14

The assumption of the truth of this theorem forms the basis of the *Śulba* method of the construction of a parallelogram having given sides inclined at a given angle, which will be described later on. It was also known in the time of the *Śatapatha Brāhmaṇa.*[1]

(*l*) *The maximum square that can be described within a circle is the one which has its corners on the circumference of the circle.*

In the *Śulba* it is sometimes necessary to draw within a circle "a square as large as possible (*yāvat saṁbhavet*);[2]" but it is not indicated how to do it. From the subsequent descriptions it, however, appears clearly, that the corners of that square are assumed to be on the periphery of the circle. The commentators explain that a side of this square will be equal to $\sqrt{2}$ times the radius of the circle. In fact, two diameters of the circle are

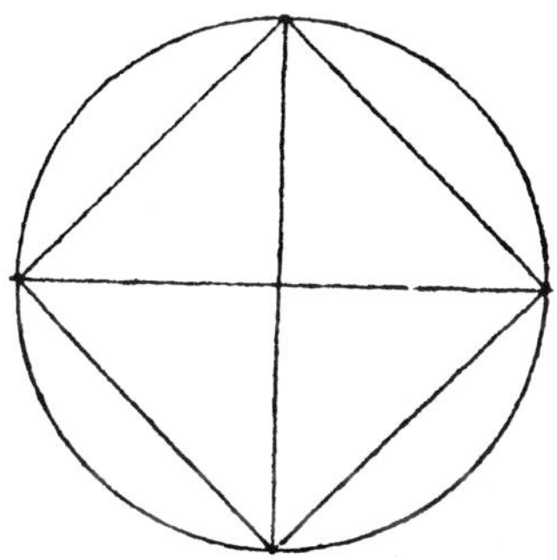

Fig. 15

drawn at right angles to each other. The figure formed by joining the end points of them, is the largest possible square within the circle.

[1] X. 2. 1. 5.

[2] *BSl*, i. 70; *ĀpSl*, vii. 10; xii. 12. Compare *Paṇḍit*, O. S., X, p. 166,

It will be noticed that every one of the above propositions is demonstrable. Branding of them as postulates raises the important question of the character of the early Hindu geometry as regards the matter of demonstration. Of course the propositions of the *Śulba* are not proved after the manner of Euclid by purely deductive reasoning. On the other hand it is not wholly empirical without any semblance of demonstration. In fact we find a kind of proof in case of the propositions of the subtraction of one square from or its addition with another square and the mensuration of an isosceles trapezium. After the enunciation as a general proposition of the theorem of the square of the diagonal, Baudhāyana observes that the truth of it will be "realised" in case of certain rational rectangles enumerated. This is really an attempt for a kind of demonstration. What is much more noteworthy in this connexion is the fact that after a description of the geometrical construction for a proposition the Śulbakāras are often found to have remarked *sa samādhiḥ* or "This is the construction."[1] The significance of such an observation is obvious. It emphasizes that the construction which was required to be made, has been thus actually made, and indeed corresponds to the expression *Quod Erat Faciendum* (or "What it was required to do") occurring at the end of a proposition of Euclid's *Elements*. Further it discloses a rational and demonstrative attitude of the mind of the early Hindu geometer. With reference to a similar remark occurring in the works of the celebrated Hindu mathematician of the twelfth century of the Christian era, Bhāskara II, Hankel observed: "The small word 'see' along with the figure together with the necessary auxiliary lines supplies the Brāhmaṇas with the

[1] *ĀpŚl*, i. 2; *KŚl*, ii. 6; iii. 13; etc.

'proof of the Greeks' concluding with solemn words 'what was to be proved.' All that a practised mind could recognise by means of assiduous consideration of a figure was admitted as certain."[1]

---

[1] H. Hankel, *Zur Geschichte der Mathematik in alterthum und mittelalter*, Leipzig, 1874, pp. 205 f.

CHAPTER V

CONSTRUCTIONS

To draw a straight line at right angles to a given straight line.

(a) Suppose that the given straight line runs east-to-west. On it fix two poles at an arbitrary distance apart, says Kātyāyana. Then

"Increase a cord of length equal to the distance between them (poles) by itself and make two ties at the ends. Then having fastened the two ties at the two poles, stretch the cord by its middle point towards the south and fix a pole at the place reached by the point. Proceed similarly on the north. It (the line joining these two poles) is the north-to-south line."[1]

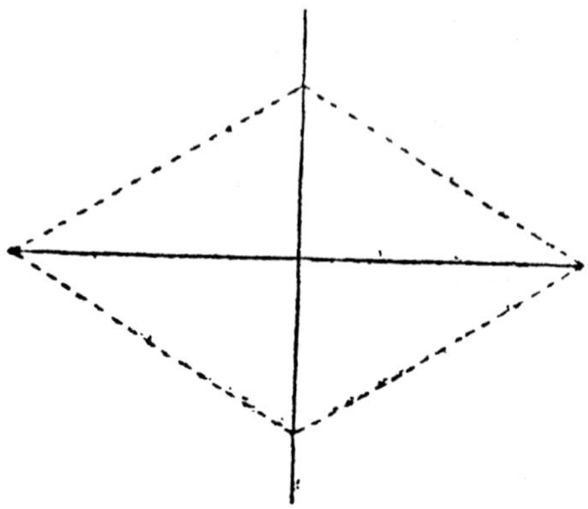

Fig. 16

1 KŚl, i. 3.

But the more common and at the same time the oldest Hindu method of drawing a straight line at right angles to another is as follows:

(b) Take two points on the given straight line. Describe two circles with their centres at these two points and their radii equal to the distance between them. The

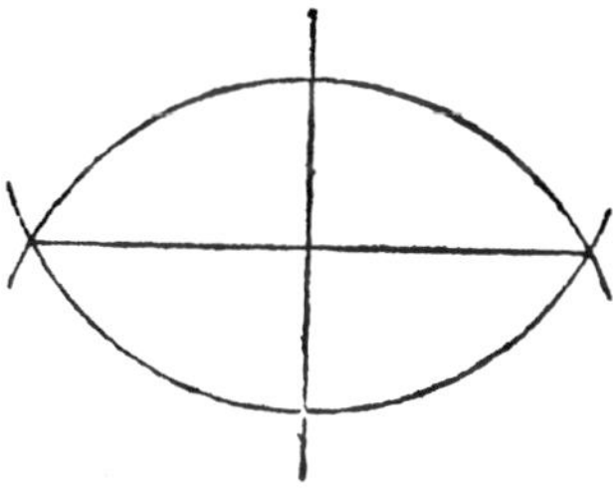

Fig. 17

line joining the points of intersection of the two circles is perpendicular to the given line.

To draw a straight line at right angles to a given straight line from a given point on it.

It should be observed that there is no particular rule for this construction in any *Śulba-sūtra* except perhaps the *Kātyāyana Śulba Pariśiṣṭa.* But it appears from the descriptions of other constructions that more than one device were used to be adopted for that purpose. The earliest of these methods is as follows:

(a) Take two points (B, C) on the given straight line (BC) at equal distance from the given point (A). With centre B and radius BC describe a circle. Similarly with

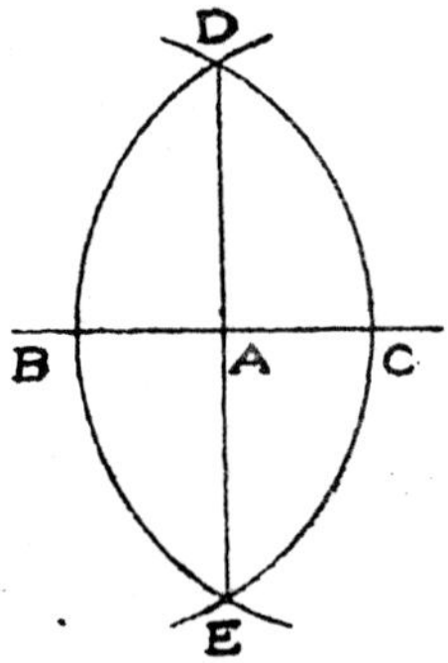

Fig. 18

centre C and radius CB describe another circle. Let D and E be the points of intersection of these two circles. Join AD or AE or DE. Then this straight line will be at right angles to the given straight line BC at A.

(*b*) On the given straight line (BC), fix two poles (B,C) equally distant from the pole at the given point A.

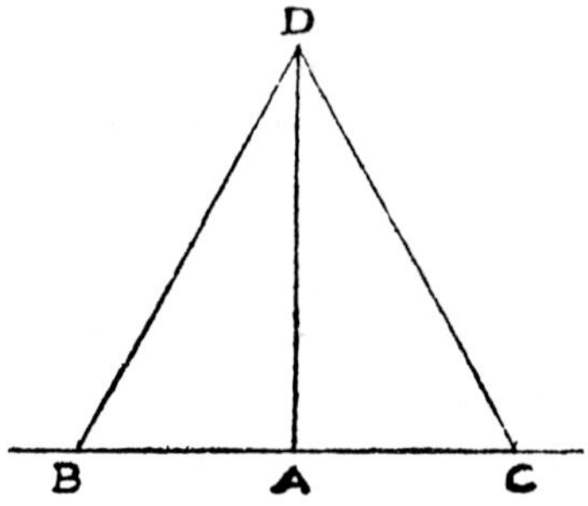

Fig. 19

Take a cord twice as long as BC. Make a tie at each of its ends and a mark at the middle. Fasten the two ties at the poles B, C and stretch the cord towards the side having taken it by the middle mark. Fix a pole D at the point reached by the mark. Join DA. It is the required straight line which is at right angles to BC at the given point A.

(c) Fix a pole (B) at a certain distance from the pole at the given point (A) on the given straight line (AB).

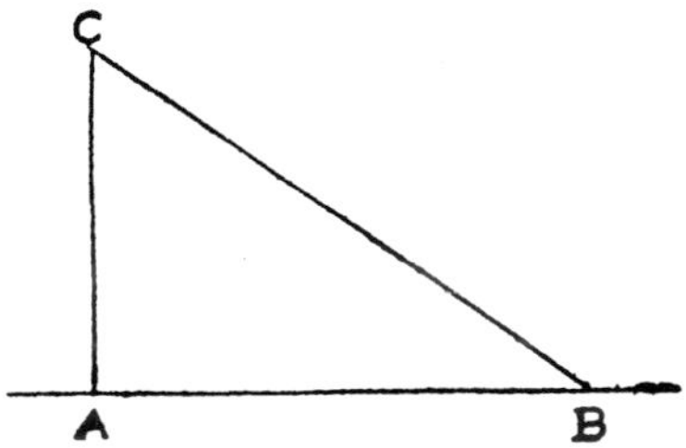

Fig. 20

Take a cord of suitable length. Make a tie at both ends and a mark at a proper point in it. Having fastened the ties at the poles (A,B), stretch the cord sidewise by the mark and fix a pole (C) at the point reached by it. Then AC is the required straight line.

To construct a square having a given side.

Method 1:

" In a bamboo-rod, make two holes (A,B) as much apart as the height of the sacrificer with uplifted arms[1] and a third hole (C) mid-way between them. Place the bamboo-rod on the east-to-west line and fix poles in the holes (beginning) from the western extremity of the sacrificial place. Then freeing the two poles (C,B) on the west, describe a circle (by rotating the bamboo) south-east-wise by the hole at the (opposite) end. Then unloosening the eastern hole and fixing the hole in the west (in

[1] The square to be constructed is to have, in the present case, a side of that length.

its original position), describe another circle south-west-wise by the hole at the opposite end. Now release the bamboo (completely); fix again an extreme hole at the middle pole (*C*) ; place it towards the south over the point of intersection of the two circles and fix a pole at the point (*F*) reached by the outermost hole. Then fix at this pole the middle hole of the bamboo and having laid it along the extreme outer edges of the two circles,[1] fix two poles (*E*,*D*) at the two (outermost) holes. It (the figure thus described, *ABDE*) is a square (having a side) of one *puruṣa*."[2] (Fig. 21.)

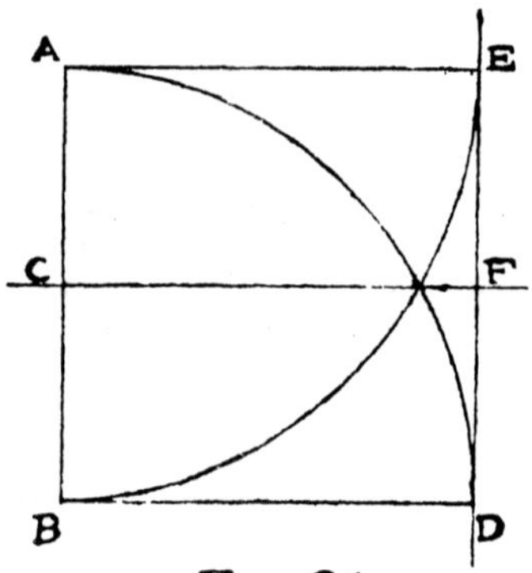

Fig. 21

Method II:

"If you wish to construct a square, take a cord as long as its side is desired to be ; make a tie at both ends and a mark at the middle. Then having drawn a line (east-to-west) of the desired length, fix a pole at its middle. Fasten the two ties at this pole and describe a circle with the mark. Now fix poles at the both ends of the diameter (running east-to-west). Having fastened one tie at the eastern pole, describe a circle with the other

[1] That is, the bamboo should be laid tangentially to both the circles and the poles are to mark the points of contact.

[2] *ĀpŚl*, viii (8-10)-xi (1).

tie. Describe a similar circle about the western pole. On joining the points of intersection of the circles, the second (*i.e.*, north-to-south) diameter will be found. Fix two poles at the extremeties of this diameter. Now, having fastened both ties at the eastern pole, describe a circle with the mark. Similarly describe circles about the southern, western and northern poles. The exterior points of intersection of these circles will determine the square."[1] (Fig. 22.)

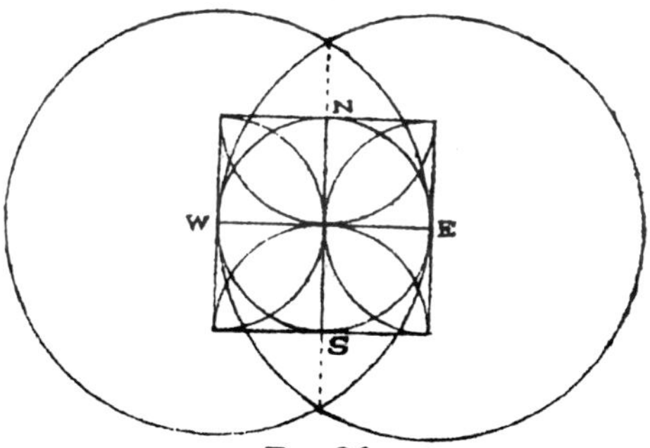

Fig. 22

Method III:

" Take a cord as long as the measure (to be given to the side of the square) ; make a tie at both ends and a mark at the middle of itself and of its two halves. Stretch out this cord along the east-west line and fix poles at the ties and marks. Then having fastened the ties at the two poles of outer marks, stretch the cord towards the south having taken it by the middle mark and make a point there. Now fasten both the ties at the middle pole and stretch the cord towards the south by the middle mark over this point and fix a pole at the place reached. Fasten one tie at this pole, another tie at the easternmost

1 *BSl*, i. 22-28.

pole, and stretch out the cord having taken it by the middle mark ; thus will be obtained the south-eastern corner of the square (required). Then freeing the tie from the easternmost pole, fasten it to the westernmost pole and again stretch the cord by the middle mark ; thus the south-western corner will be determined. Similarly can be determind the north-eastern and north-western corners of the square."[1] (Fig. 23.)

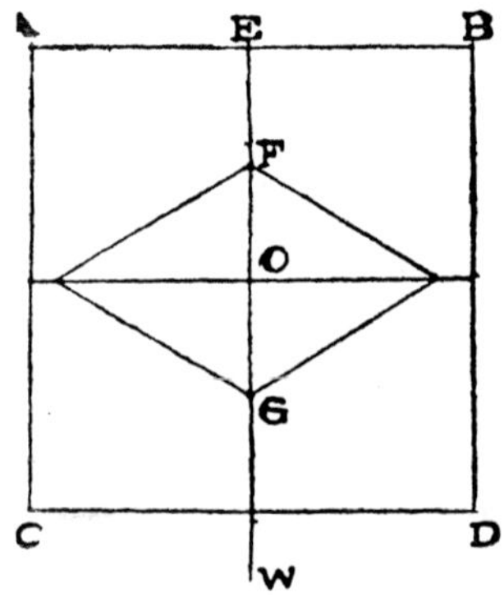

Fig. 23.

Method IV :

" Take a cord two times the measure of the given side of the square; make a tie at both ends and a mark at the middle. With one half of this cord measure the east-to-west (breadth) of the square. In the other half, make a mark at a distance (from the western end) less by its one-fourth. Let this mark be called *nyañchana*. Make another mark at the middle of that half for the purpose of (determining) the eastern corners. Having fastened the two ties at the two extremities of the east-to-west breadth, stretch the cord towards the south by the *nyañchana* mark. Thus the two eastern and two

[1] *ĀpŚl*, i. 7.

western corners of the square should be constructed by the middle mark of the other half of the cord."[1] (Fig. 24.)

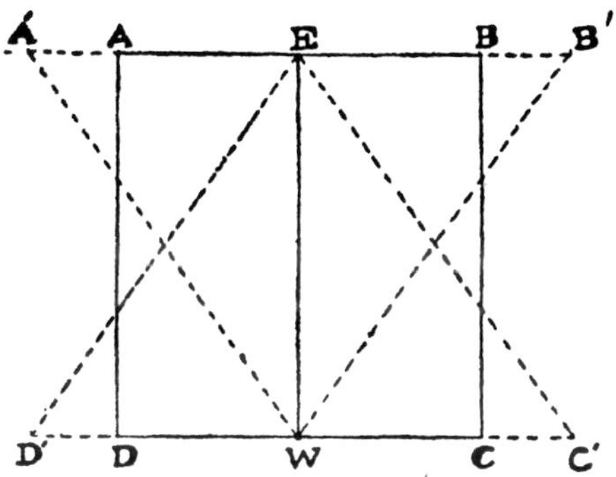

Fig. 24

$EW = a$, the given side; $AE = EB = DW = WC = a/2$; $A'E = EB' = D'W = WC' = 3a/4$; $ED' = EC' = WA' = WB' = 5a/4$.

Method V :

"Add to a cord as long as the given side its half and make a mark at a distance (from the other end of the added portion) less by its sixth part. Fasten the ends of the (increased) cord at the extremities of the east-west line and stretch it towards the south having taken by the mark and put a sign at the point reached by it. Do similarly on the north and again on both sides after interchanging the ends of the cord. This is the construction." [2] (Fig. 25.)

1 *BŚl*, i. 29-35.

2 *ĀpŚl*, i. 2.

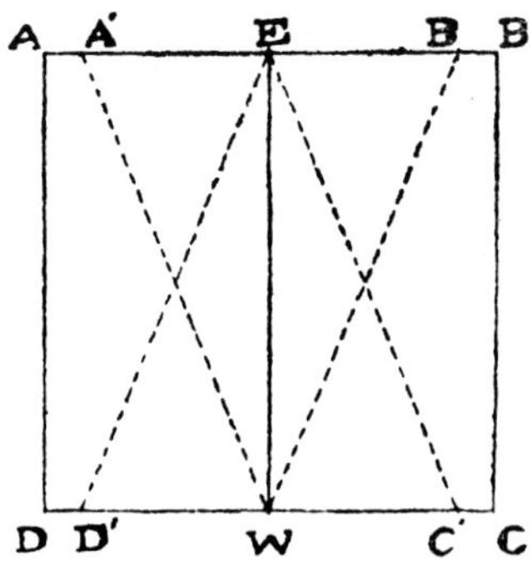

Fig. 25

$EW=a$, the given side; $AE=EB=DW=WC=a/2$; $A'E=EB'=D'W=WC'=5a/12$; $ED'=EC'=WA'=WB'=13a/12$.

The Method I is described in full by Āpastamba.[1] It is noted and partly described by Baudhāyana.[2] This seems to be the oldest Hindu method for the construction of a square on a given straight line. For, the practice of the measurement of the Fire-altar with the bamboo-rod is mentioned as early as the *Taittirīya Saṁhitā* (c. 3000 B.C.)[3] and indeed reappears in almost all the early

[1] *Loc. cit.*

[2] *BŚl*, iii. 13ff.

[3] *TS*, v.2.5. 1ff.

In the early *Saṁhitā* and *Brāhmaṇa* is found a mythological relation between the *Agni* and the *Veṇu* ("bamboo-rod"). Thus *Taittirīya Saṁhitā* observes, "He metes with a bamboo; the bamboo is connected with Agni; (verily it serves) to unite him with his birth place" (v.2.5.2). The connection has been narrated in that work thus: "Agni went away from the gods; he entered the reed; he resorted to the hole which is formed by the perforation of the reed." This mythology reappears in the *Maitrāyaṇī Saṁhitā* (iii. 2. 4) and *Śatapatha Brāhmaṇa*. The latter describes, "Agni went away from the gods; he entered into a reed, whence it is hollow, and whence inside it is,

Saṁhitās and *Brāhmaṇas*.[1] It is found to have been gradually replaced by the measurement with a cord which was introduced about the time of the *Śatapatha Brāhmaṇa* (*c.* 2000 B.C.).[2]

The Method II occurs only in the *Baudhāyana Śulba*. It is clearly based on the previous method and is indeed a combination of four operations by it.[3]

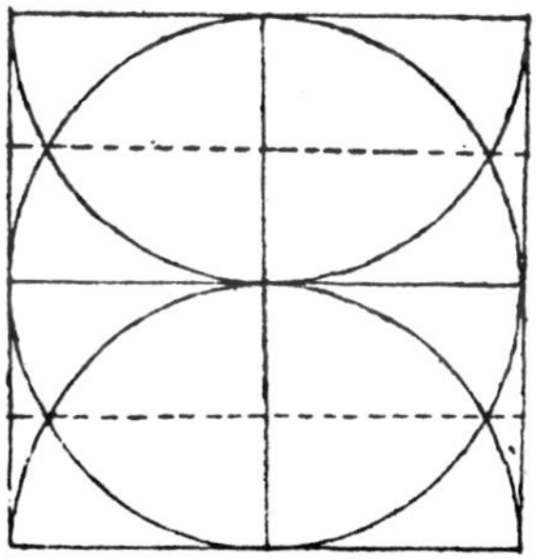

Fig. 26

The Method III is taught by Āpastamba[4] and Manu.[5] The latter emphasizes that it should be exclusively used in all cases of the construction of a square on a given

as it were, smoke-tinged" (vi.3.1.26); again "Agni went away from the gods. He entered into a bamboo-stem; whence that is hollow. On both sides he made himself those fences, the knots, so as not to be found out; and wherever he burnt through, those spots came to be" (vi.3.1.31).

1 *MaiS*, iii.2-4; *KṭŚ*, xx.3-4; *KapS*, xxxii.5.6; *ŚBr*.

2 *SBr*, x. 2.3.8 ff.

3 Compare Bürk, *ZDMG*, lv, p. 547.

4 *Loc. cit.*

5 *MāŚl*, i. 14-21; compare also vii. 7 ff.

straight line.[1] This method is also applied by Kātyāyana.[2] It may be noted that the cord used in the course of the construction is called in the *Mānava Śulba* by the technical name *pañcāṅgi* ("five-jointed"), because it has five (*pañca*) joints (*aṅga*), *viz.*, two ties and three marks. The rope has been sometimes called *pātinī* ("that which is laid out") because measurements are made by laying (*pāta*) it out on the ground.

The Method IV is given by Baudhāyana,[3] Āpastamba,[4] Kātyāyana,[5] Manu[6] and Maitrāyaṇa; and the Method V by Āpastamba[7] and Kātyāyana.[8] It is also taught by Baudhāyana.[9] He would, however, restrict its application to the construction of a rectangle.

To construct a rectangle of given sides.

"If you wish to construct a rectangle, fix on the ground two poles as much apart as you wish (the length to be). On either sides (before and behind) of each of these poles, fix two other poles at equal distances from it. Take a cord as much as the breadth (of the rectangle); make a tie at both ends and a mark at the middle. Having fastened the two ties at the two poles about the eastern pole, stretch the cord towards the south by the mark and put a sign (where the mark touches the ground). Then fasten the two ties at the middle pole and again draw the

1 *Ibid*, vii. 7.
2 *KŚr*, xvi. 8. 1-20.
3 *Loc. cit.*
4 *ĀpŚl*, i. 3.
5 *KŚl*, i. 12-3; compare also *KŚlP*, 16
6 *MāŚl*, iii. 5 ff.
7 *Loc. cit.*
8 *KŚl*, i. 14-5.
9 *BŚl*, i. 42-4.

cord over the sign towards the south by the mark and fix a pole at the mark. That is the south-east corner (of the rectangle). Thereby is explained (how to determine) the north-east corner and also the two western corners." [1]

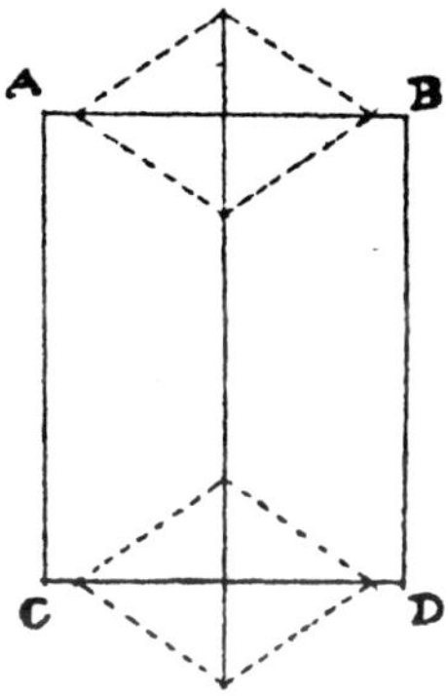

Fig. 27

This method is alike in principle to that employed by Āpastamba and Manu for the construction of a square on a given straight line (*Method III*). Still Baudhāyana would restrict its application to the construction of a rectangle having given sides and of an isosceles trapezium having given face, base and altitude.

To construct an isosceles trapezium of a given altitude, face, and base.

To construct a trapezium whose altitude, face and base are given, Baudhāyana [2] follows a method similar to the one adopted by him for the construction of a rectangle of given sides. Only cords of given different measures are employed for fixing the extremities of the face and base.

[1] *BŚl*, i.36-40.

[2] *BŚl*, i.41.

The methods suggested by Āpastamba for the same purpose will be understood from those adopted by him for the construction of the *Mahāvedi*. The shape of the *Mahāvedi* is prescribed by tradition to be an isosceles trapezium whose altitude is 36 *pada* (or *prakrama*), face 24 (units) and base 30 (units). Āpastamba gives four methods for its construction. All of them are in principle the same. It is to draw a straight line through both the extremities of another straight line equal to the given altitude and at right angles to it. Along these straight lines and on other sides of the altitude are measured lengths equal to half the given lengths of the face and the base. And thus the isosceles trapezium is drawn. Āpastamba distinguishes between his different methods as *Ekarajjvāviharaṇa* ("The method of construction with one cord") and *Dvirajjvāviharaṇa* ("The method of construction with two cords").

Method I :

"Add to a cord of 36 (*pada* or *prakrama*) 18 and make a mark at 12 and a mark at 15 from its western end. Having fastened the ends of the cord to (the two poles at) the two extremities of the east-west line (of 36 *pada*) stretch it towards the south by taking by the mark at 15

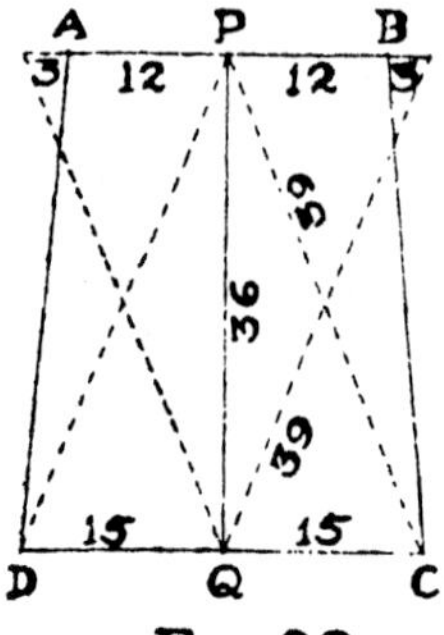

Fig. 28

and fix a pole (at the point reached by the mark); (proceed) similarly towards the north. These poles are the two western corners of the *vedi*. For (determining) the two eastern corners, interchange (the ends of the cord) and then stretching it towards the south by the mark at 15, fix a pole at (the point reached by) the mark at 12; (proceed) similarly towards the north. These are the two eastern corners. This is the method of construction with one cord."[1] (Fig. 28.)

Method II:

"The diagonal of a rectangle whose sides are 3 and 4 (*pada* or *prakrama*) is 5. With these increased by three times themselves (are determined) the two eastern corners of the *vedi*. With them increased by four times themselves (are fixed) the two western corners." [2] (Fig. 29.)

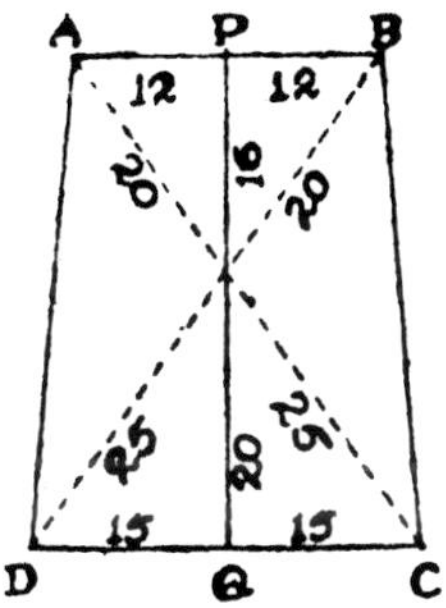

Fig. 29

$3^3 + 4^2 = 5^2$

$3 + 3.3 = 12$
$4 + 3.4 = 16$
$5 + 3.5 = 20$

$12^2 + 16^2 = 20^2$

$3 + 4.3 = 15$
$4 + 4.4 = 20$
$5 + 4.5 = 25$

$15^2 + 20^2 = 25^2$

Method III:

"The diagonal of a rectangle whose sides are 5 and 12, is 13. With them the eastern corners of the *vedi* (are

ĀpSl, v.2.
ĀpSl, v.3.

determined); and with them increased by twice themselves the western corners (are fixed)"[1] (Fig. 30.)

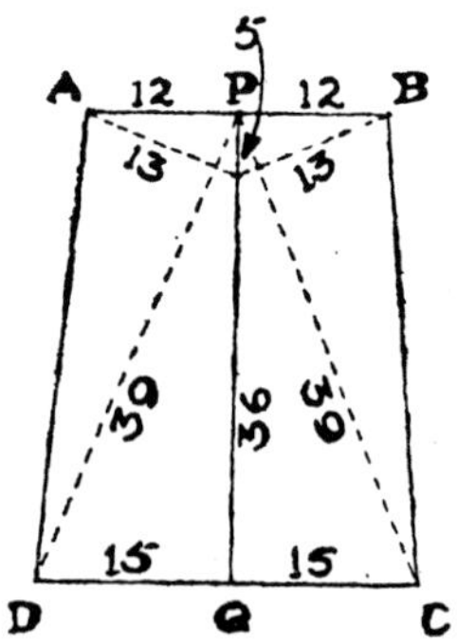

$$5^2+12^2=13^2$$

$$5+2.5=15$$
$$12+2.12=36$$
$$13+2.13=39$$

$$15^2+36^2=39^2$$

Fig. 30

Method IV :

" The diagonal of a rectangle whose sides are 8 and 15, is 17; with these cords, the two western corners of the *vedi* (are measured). The diagonal of a rectangle of sides 12 and 35 is 37 ; with these cords, (are measured) the eastern corners."[2] (Fig. 31.)

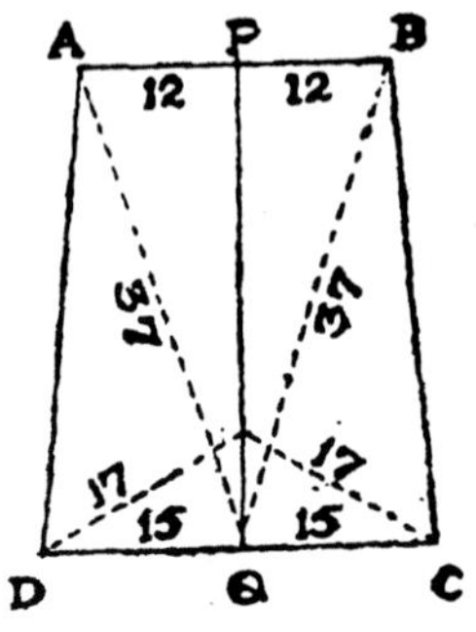

$$8^2+15^2=17^2$$
$$12^2+35^2=37^2$$

Fig. 31

[1] *Āpśl*, v.4.
[2] *Āpśl*, v.5.

Āpastamba observes: "These are the known methods of construction of the (*Saumikī*) *vedi*." [1]

To construct a parallelogram having given sides at a given inclination.

For the proper construction of an altar of prescribed size and shape, it was sometimes necessary to make bricks of the shape of a parallelogram having given sides at a given inclination. For instance, Āpastamba says:

"Make a class of bricks, one-fifth of a puruṣa long and one-ninth of a puruṣa broad, (the two sides being) inclined (*nata*) suitably so as to fit (*yathā-yagaṃ*)."[2]

The reference in the end of the rule is to the bent of the wing of the falcon-shaped altar where these bricks shall have to be used. Thibaut observes: "By 'nata, bent' the sūtrakāra means to indicate the sides of the brick bo not form rectangles. The shape of the brick is rhomdoidical, the angles, which the sides form with each other are the same which the wings of the *śyena* form with the body."[3]

Āpastamba does not expressly teach us how to construct a parallelogram of that shape. But it could be inferred from the method laid down for the construction of another parallelogram. He says,

"(Construct) bricks for the wing with four sides: two sides one-fourth puruṣa each and two sides of one-seventh puruṣa."[4]

By the expression *pakṣeṣṭaka* ("brick for the wing") is implied that the inclination between the two contiguous sides of these bricks must be the same as the bending of

[1] "Etāvanti jñeyāni vedi-viharanāni bhavanti"—*ĀpŚl*, v.6.

[2] *ĀpŚl*, xvi.2.

[3] Thibaut, *Śulvasūtras*, p. 31.

[4] *ĀpŚl*, xix.5.

the wings where they are to be used. Now the bending of a wing has been defined thus :[1] Describe a rectangle *ABCD* of breadth (*AD*) equal to 1 puruṣa and length (*AB*)

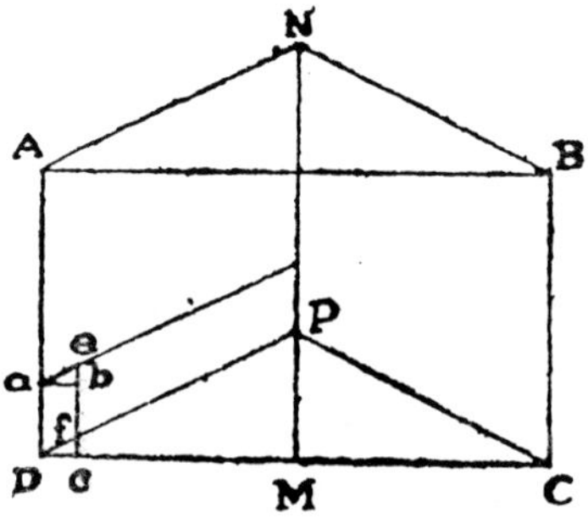

Fig. 32

$2\frac{1}{16}$ puruṣa. Draw *MN* bisecting *DC* and at right angles to it. Draw *DP* one puruṣa long meeting *MN* at *P*. Cut off *PN* equal to *DP*. Draw *NA*, *NB*, *PD*, *PC*. This will give the bending of the wing.

Āpastamba teaches the following method for construction of the above *pakṣeṣṭaka* :

" Construct a rectangle (*abcd*) of length (*aD*) equal to one-fourth of a puruṣa and breadth (*Dc*) equal to one-seventh of the side (*DM*) of (half) the wing; by the slanting side (*DP*) the mouth of this rectangle should be changed so that the sides (*Df* and *ae*) might be bent to the one-seventh of the slanting side of the wing (*DP*)." [2]

Thus we have the required parallelogram *acfD* whose sides *aD*, *fc* are equal to one-fourth of a puruṣa each and sides *fD*, *ae* are each equal to one-seventh of a puruṣa and they being inclined as specified.

[1] *Cf. ibid*, xviii. 6-8.

[2] *ĀpŚl*, xix.8.

It should be particularly noticed that in the above the inclination between the two sides of the parallelogram to be constructed or between any two lines in general is not stated in terms of angular units (though the conception of such is not wanting in Hindu astronomy) but in terms of the relation between the sides and diagonals of a certain rectangle along which the two given lines would lie.

Āpastamba gives also a slightly different method of constructing a parallelogram. He says:

"Draw a rectangle one-fifth of a puruṣa long from east to west, and one-tenth of a puruṣa broad ; to the south as well as to the north of it draw another (rectangle of the same size). Draw the diagonals of them passing through their south-western corners."[1]

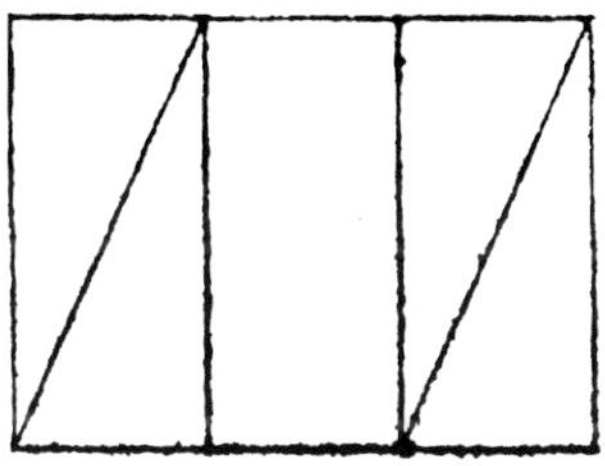

Fig. 33

From this we can easily guess the method that was generally followed in the *Śulba* for the construction of a parallelogram having given sides at a given inclination. Let A, B represent the given sides and the inclination be the same as that between the diagonal PR and the side

[1] *Ibid*, xvi.8.

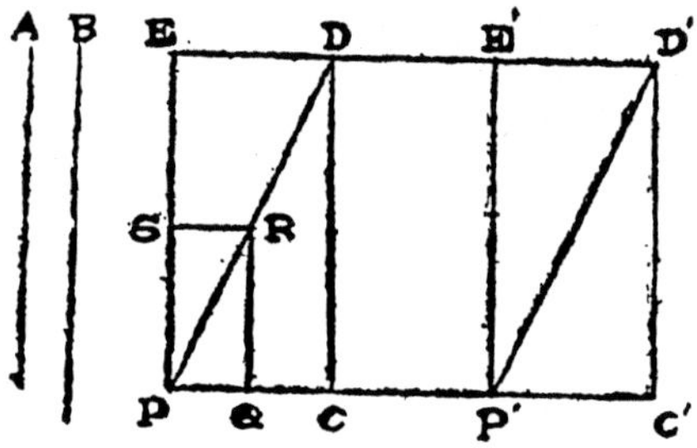

Fig. 34

PQ of the rectangle $SRQP$. Now draw the rectangle $EDCP$ similar to the given rectangle such that its diagonal PD be equal to the given side B. Then draw the rectangle $DE'P'C$ so that its side CP' be equal to the difference between A and PC. Now draw the rectangle $E'D'C'P'$ equal to the rectangle $EDCP$. Join $D'P'$. Then $DD'P'P$ is the required parallelogram. The Fig. 34 answers to the case when $A > PC$. In case $A < PC$, the rectangle $DE'P'C$

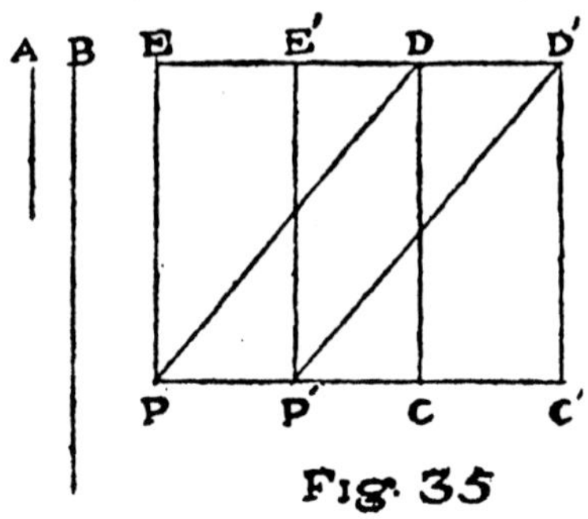

Fig. 35

should be drawn so as to overlap the rectangle $EDCP$. (Fig. 35.)

CHAPTER VI

COMBINATION OF AREAS

To draw a square equivalent to n times a given square.

The solution of this proposition practically depends upon the construction of squares and rectangles having given sides. By the so-called Pythagorean Theorem, the square described on the diagonal of the given square will have an area twice as much. To draw a square thrice as much as the given square, the rule is:

"(Construct a rectangle whose) breadth will be the measure (of a side of the given square) and length its double-producer (*i.e.*, diagonal). The diagonal of that rectangle is the treble-producer."[1]

By repeating the operation and constructing each time a rectangle whose one side will be equal to a side of the given square and another side equal to the diagonal of the rectangle constructed at the preceding stage, we shall finally arrive at one, the square on whose diagonal will be equivalent to n times the given square. This method is taught by almost all the writers.

Oftentimes the process can be much shortened by a skilful device. For instance, if n happen to be a square number, equal to p^2, say, then the desired result will be obtained by drawing the square on a straight line p times a side of the given square. Thus Kātyāyana observes:

"Twice the measure (of a side of a given square) is (its) fourfold-producer; thrice the measure is the ninefold-producer; four times the measure is the

[1] *BSl*, i, 46; *ĀpSl*, ii. 2; *KŚl*, ii. 14.

sixteenfold-producer. As many units of a measure as are in a cord, so many rows (or series) of squares (of that measure) there will be in a square on that cord as a side. Combine them."[1]

If n is not a square number, simplification can be made by expressing it, when possible, as the sum of two square numbers. Thus if

$$n=p^2+q^2$$

where p, q are rational quantities, then if x denote a side of the required square and c a side of the given square, we shall have

$$x^2=(pc)^2+(qc)^2=(p^2+q^2)c^2=nc^2.$$

So that with one construction only, we shall get a square n times the given square. Kātyāyana has given some instances of this kind. He says:

"(If a rectangle be drawn with) one pada (a unit of linear measure) as the breadth and three padas as the length, its diagonal will be tenfold-producer. (If a rectangle be drawn with) two padas as the breadth and six padas as the length, its diagonal will be fortyfold-producer."[2]

The first of these instances occurs also in the *Mānava Śulba*.[3] Similar instances are also found in other works.

A simplification is also possible even if n is expressible as a multiple of a square number, though not as the sum of two square numbers.

Kātyāyana gives another very elegant and simple method of finding a square equal to the sum of a number of other squares of the same size. He says:

"As many squares (of equal size) as you wish to combine into one, the transverse line will be (equal to) one

[1] *KŚl*, iii. 6-9 ; compare also *ĀpŚl*, iii. 6-7.

[2] *KŚl*, ii. 8-9.

[3] *MāŚl*, v. 4-5.

less than that ; twice a side will be (equal to) one more than that ; (thus) form a triangle. Its arrow (*i.e.*, altitude) will do that."[1]

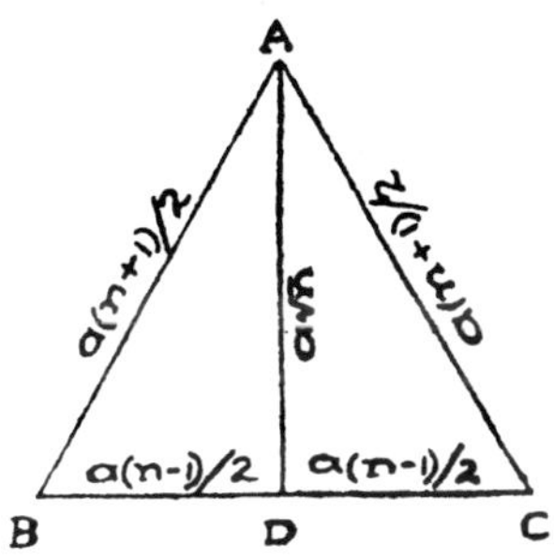

Fig. 36

That is, if n be the number of equal squares to be combined together into one, form the triangle ABC whose base BC is of length $(n-1)$ times a side of a square and twice the sides AB and AC of which are severally equal to $(n+1)$ times a side of a square. The method of drawing the triangle, which will be consistent with the geometrical methods of the *Śulba* is this : Draw the line BC of length $(n-1)$ times a side of a square. Fix two poles at B and C. Take a cord of length $(n+1)$ times a side of a square. Fasten its two ends at the two poles and stretch the cord sidewise having taken it by the middle point. Let A be the point reached. Bisect BC at D and join AD. Then the square on AD will be equivalent to

[1] "यावत् प्रमाणानि समचतुरस्राण्येकीकर्तुं चिकीर्षेदेकोनानि तानि भवन्ति तिर्यक् द्विगुणान्येकत एकाधिकानि त्र्यश्रिर्भवति, तस्येषुस्तत्करोति ।" —*KŚl*, vi. 5. Compare also the *Pariśiṣṭa*, verses 40-1.

the sum of n given squares. For

$$AD^2 = AC^2 - DC^2,$$

$$= \left(\frac{n+1}{2}\right)^2 a^2 - \left(\frac{n-1}{2}\right)^2 a^2,$$

$$= na^2.$$

To draw a square equivalent to the nth part of a given square.

After describing the method of drawing a square equivalent to 3 times a given square, Baudhāyana observes:

" Thereby is explained the generator of the third part (*tṛtīyakaraṇī* of the square). It is the ninth part of the area."[1]

Similar remarks occur also in the works of Āpastamba and Kātyāyana. The former says,

" Thereby is explained the generator of the third part. Division into nine (parts)." [2]

and the latter,

" Thereby is explained the generator of the third part. Division of the given measure into nine parts." [3]

The commentators disagree about the method actually implied in the above rules. According to some,[4] it is this: Find a square equivalent to three times the given square. Then divide a side of this square into three equal parts. A square drawn on any of these parts will be equivalent to one-third of the given square. For, on drawing parallel lines through the points of division, the second square will be divided into nine (square) parts. Hence each part (square) is equal to one-ninth of it and so

1 *BŚl*, i. 47.

2 *ApŚl*, ii. 3.

3 *KŚl*, ii. 15-6.

4 For instance, Kapardisvāmī, Sundararāja, and Rāma.

to one-third of the given square. According to others,[1] the method is this: Divide the given square into nine equal squares. Combine three of these squares into one. And it will be equivalent to the third part of the given area. Some[2] are of opinion that both the methods are implied by the text.

Both the interpretations are valid inasmuch as they produce the correct result. Thibaut prefers the first interpretation as it preserves in a better way the connexion of the above rule with those just preceding. But the Sulbakāras appear to have implied both the methods. Kātyāyana's rules following those noted above are:

"The third part of the side (of the given square) produces the ninth part of it. Three of these ninth parts (on combination) will give the generator of the third part (of the given square)."[3]

Again this is the method adopted by Baudhāyana in constructing the *Paitṛkī-vedi* which is square in shape and is equivalent to one-third the square on a side 18 padas long.[4] Further he has expressly adopted both the methods in measuring the *Sautrāmaṇīkī-vedi*.[5]

Proceeding in the same way we can find in general a square equivalent to the nth part of a given square. *Either*: First find a square equivalent to n times the given square. The square on the nth part of a side of this square will be equivalent to the nth part of the given square. *Or* Divide a side of the given square into n equal parts. On drawing parallel lines through the points of division, the given square will be divided into n^2 equal

[1] *E.g.*, Dvārakānātha Yajvā.

[2] Karavindasvāmī.

[3] *KŚl*, ii. 17-8.

[4] *Vide infra*, p. 98f.

[5] *BŚr*, xix. 1.

square parts. The square combining n of these elementary squares will be equivalent to the nth part of the given square.

The whole process will, of course, be much simplified if n be a square number, equal to p^2, say. For in that case we shall have simply to divide a side of the given square into p equal parts. The square having one of these parts as a side will be equal to nth part of the given square. Kātyāyana furnishes us with some instances of this kind:

"By means of half the measure (of the side of a given square) is obtained a square equivalent to the fourth part (of the given square); by one-third the measure is obtained (a square equivalent to) the ninth part; by one-fourth the measure is obtained (a square equivalent to) the sixteenth part."[1]

Similar instances are also given by Āpastamba.[2]

To draw a square equivalent to the sum of two different squares.

Baudhāyana gives the following method of solution of this proposition:

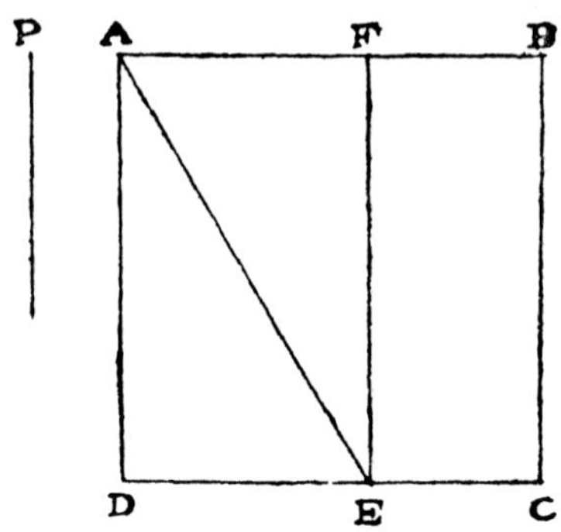

Fig. 37

[1] *KSl*, iii. 10-11.
[2] *ĀpSl*, iii. 10.

"To combine two different squares, cut off from the larger a (rectangular) portion with a side of the smaller one. The diagonal of this segment will be a side of the sum."[1]

The same method is also taught by Āpastamba [2] and Kātyāyana.[3]

Let $ABCD$ be the larger square and P a side of the smaller. Cut off AF and DE making each equal to P, and complete the rectangle $AFED$. Join AE. Then

$$AE^2 = AD^2 + DE^2 = AD^2 + P^2.$$

The proof of this proposition will be evident from the Fig. 38 which, in fact, simply represents the complete constructions taught in the *Śulba*.

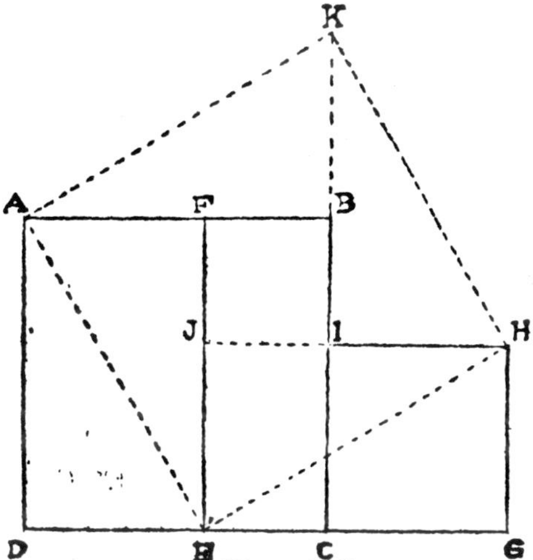

Fig. 38

$$\square ABCD + \square CGHI = \triangle ADE + \triangle AEF + \triangle EGH + \triangle EHJ + \square BIJF,$$
$$= \triangle ABK + \triangle AEF + \triangle HIK + \triangle EHJ + \square BIJF,$$
$$= \square AEHK.$$

Or $AD^2 + P^2 = AE^2$, since $CG = P$

[1] *BŚl*, i. 52. [2] *ĀpŚl*, ii. 4. [3] *KŚl*, ii. 22.

To draw a square equivalent to the difference of two different squares.

For the solution of this proposition, Baudhāyana[1] and Āpastamba[2] give the following rule:

" To deduct a square from a square, cut off from the larger a (rectangular) segment with a side of the square which is to be deducted. Then draw a longer side of this segment diagonally across to the other longer side; and where it falls (on the other side), cut off that portion. By this cut-off portion the deduction is finished."

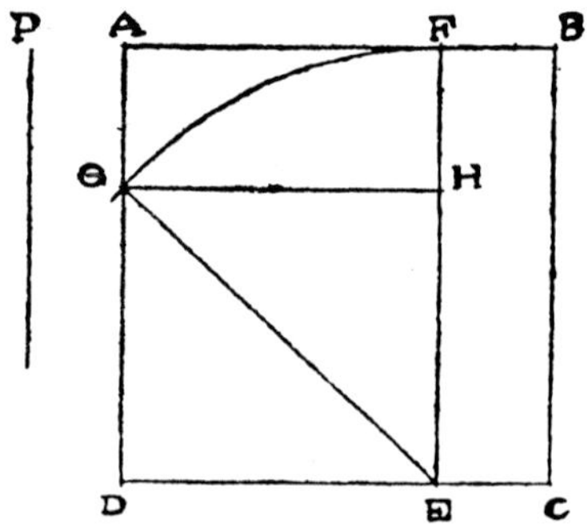

Fig. 39

Let $ABCD$ be the larger square and P a side of the smaller square to be deducted from it. Cut off AF and DE making each equal to P. Join FE. Draw FE by the extremity F so that it falls on AD at G. Join GE. Then

$$GD^2 = GE^2 - DE^2 = AD^2 - P^2.$$

This method is also taught by Kātyāyana.[3]

1 *BSl*, i. 51.
2 *ĀpSl*, ii. 5.
3 *KSl*, iii. 1.

The proof of this proposition will appear on completing the constructions indicated in the *Śulba*. The square

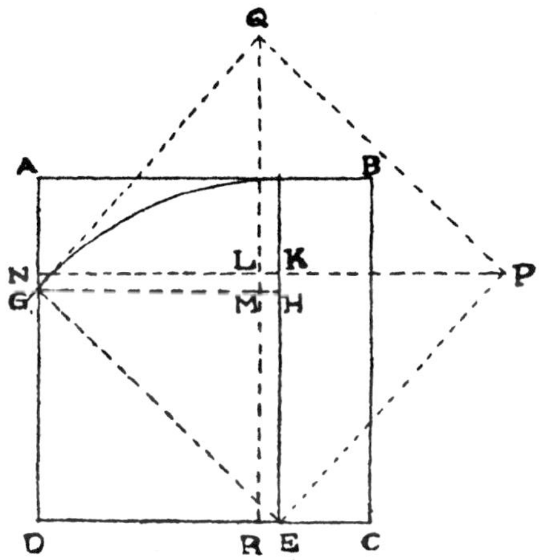

Fig. 40

$GEPQ$, which is equal to the square $ABCD$, is comprised of four rightangled triangles each equal to GMQ and of the square $HKLM$. Therefore

$$GEPQ = GDEH + DRLN + HKLM,$$
$$= GDRM + DEKN,$$
$$\therefore \quad GDRM = ABCD - DEKN,$$
$$\text{or} \quad GD^2 = AB^2 - DE.^2$$

Āpastamba has given a *demonstration* of the above rule together with an illustrative example:

" The side drawn cross becomes the diagonal (of a certain rectangle). Let it be the generator of the four-fold (of a certain square). It produces (a square equal to) both the squares which are produced separately by the cut-off portion (of the opposite side) and the other side (*i.e.*, the shorter side of the rectangular segment of the larger square or the side of the other square).

If the other side be one puruṣa, the remainder will be three square puruṣas ; it has been stated (before)."[1]

For the construction of a Fire-altar of proper size and shape, it is also necessary to know how to combine into a square figures of other kinds; *e.g.*, a square and a rectangle, two rectangles.[2] No specific rules for this purpose are found in the *Śulba-sūtra*. Hence it follows that such combinations shall have to be made with the help of the methods taught. Thus figures of every other kind are first transformed into squares and they are then combined into a square by the methods just described. This method has indeed been taught by Kātyāyana for the combination of triangles and pentagons.

To draw a square equivalent to two given triangles.

After describing a method for the transformation of an isosceles triangle or a rhombus into a square, Kātyāyana observes: " By this is explained the combination of triangles." That is, the triangles and rhombuses to be combined should first be transformed into squares severally and the sum or difference of these squares is then found by the methods already explained. The final result can, of course, be put in the shape of a square, rectangle or triangle as required.

To draw a square equivalent to two given pentagons.

Kātyāyana has also indicated a method for the combination of pentalaterals (*pañcakarṇa*) :

" By this is also explained a method for the combination of pentalaterals too. Break up a pentalateral of

[1] *ĀpŚl*, ii. 6.
[2] Cf. *ĀpŚl*, xxi. 8.

equal angles into isosceles triangles; and break up a pentalateral of unequal angles into squares."[1]

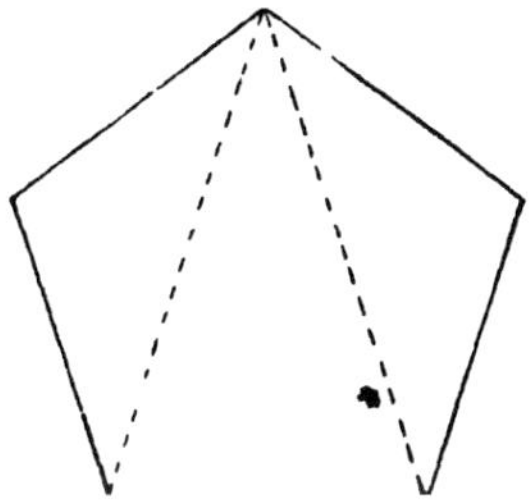

Fig. 41

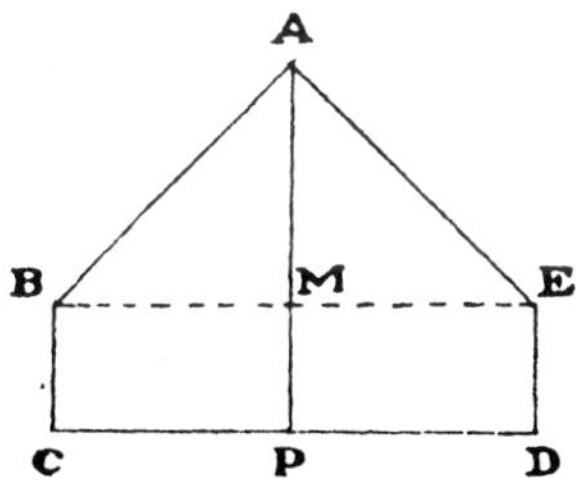

Fig. 42

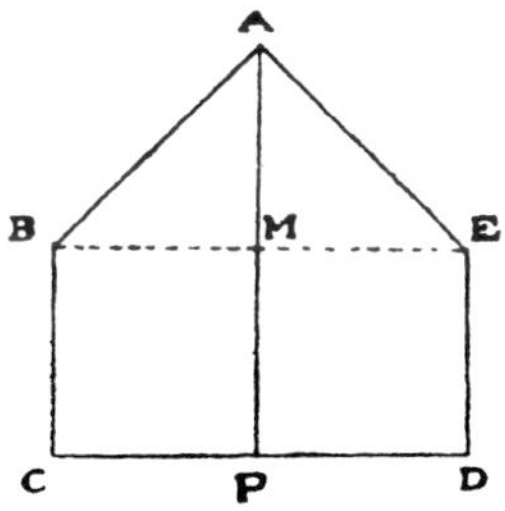

Fig. 43

The latter part of this rule appears rather to be obscure. How to break up an irregular pentagon into squares? Mahīdhara has failed to grasp the matter accurately and hence made a worse confusion.[2] I think

[1] *KŚl*, iv. 8.

[2] Mahīdhara has once explained *ekakarṇa* = "one triangle," *dvikarṇa* = "two triangles," *trikarṇa* = "three triangles" and *pañcakarṇa* = "five triangles." Later on he says *ekakarṇa* means "equal angles" and *dvikarṇa* "unequal angles." His former interpretations are certainly wrong. If Kātyāyana had indeed meant "five triangles" by the term *pañcakarṇa*, what did he mean by the rule, "Pañcakarṇānāñca praüge apacchidyaikakarṇānāṃ dvikarṇānāṃ samacaturasre apacchidya." ["Break up *pañcakarṇas* of *ekakarṇa* (variety) into isosceles triangles and break up *pañcakarṇas* of *dvikarṇa* (variety) into squares?"]

Kātyāyana has in view a pentagon of the shape $ABCDE$ in which $CP=PD=AM$ and $BC=MP=ED=CP/2$ (Fig. 42); *or* in which CP, PD, AM, BC, MP, ED are all equal (Fig. 43). We find description of bricks of these shapes in the *Śulba* of Baudhāyana [1] and they are ordinarily called *haṁsamukhī* ("of the shape of the mouth of a goose"). Obviously a pentagon of that kind can be easily transformed into two or three squares of equal size. So that by combining the transformed squares we can find a square equal to the sum or difference of two or more pentagons.

---

[1] *BŚl*, iii. 68, 288; compare also iii. 291-2.

CHAPTER VII

TRANSFORMATION OF AREAS

To transform a rectangle into a square.

For this purpose Baudhāyana gives the following rule:

"If you wish to transform a rectangle into a square, make its breadth as the side of a square; divide the remainder into two parts and changing the place (of the farther one of them) and inverting, add it on the other side of the square. Then adding a (square) portion, fill up that (the empty space in the corner). It has been taught (before) how to deduct it (the added square from the full square thus formed)."[1]

Let it be required to change the rectangle $ABDC$ into a square.

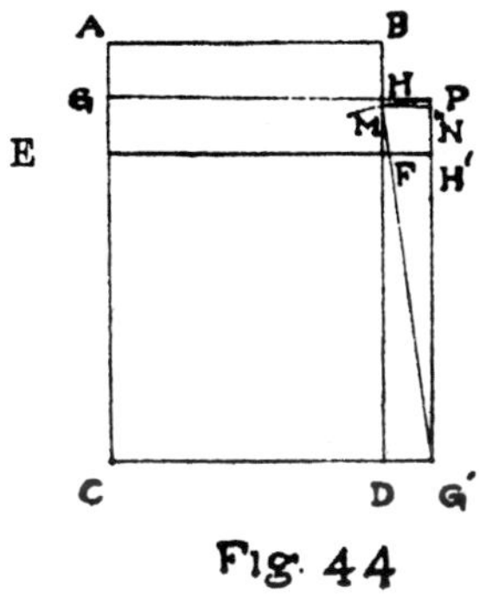

Fig. 44

From the longer side AC cut off a portion CE equal to the breadth CD of the rectangle. Complete the square $CDFE$. Divide the remaining part $ABFE$ of the given rectangle into two halves by the line GH. Take the remoter half $ABHG$ and place it after inversion on the other side of the square $CDFE$ in the position $DG'H'F$. Complete

[1] *BŚl*, i. 58.

the square $CGPG'$, by adding the portion $HPH'F$. The given rectangle is easily found to be equal to the difference of the two squares $GPG'C$ and $HPH'F$. This difference can be found by the method taught before. That is, draw a circle with centre G' and radius $G'P$ cutting DH at M. Draw MN perpendicular to $G'P$. Then

$$G'N^2 = G'M^2 - MN^2 = G'P^2 - HP^2$$

so that $G'N$ is the side of the square which is equivalent to the given rectangle $ABDC$.

The same method is taught also by Āpastamba[1] and Kātyāyana.[2] The latter is a little more explicit than the other. He says:

"If you wish to transform a rectangle into a square, cut off (from the rectangle a square portion) with its shorter side. Divide the other portion into two parts. Take the farther portion on the east and add it on the south (of the square portion). Complete the (square) figure by introducing a (small square) piece. The method for its deduction has been taught."

Kātyāyana suggests a different procedure for the transformation of a rectangle whose length much exceeds its breadth:

"If (the rectangle be) very long, cut it again and again (into squares) by the breadth ; combine these squares into one square ; add to this the remaining portion (of the rectangle) after transforming it suitably." [3]

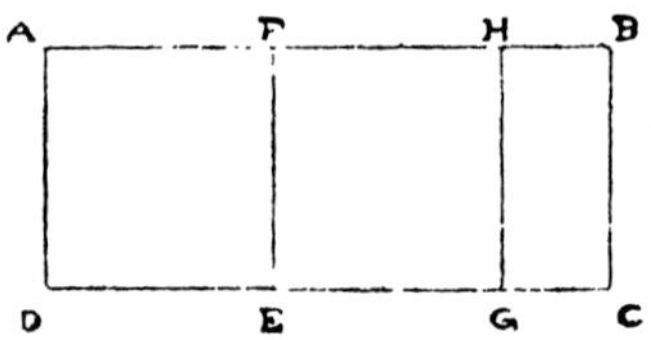

Fig. 45

[1] *ĀpŚl*, ii. 7. [2] *KŚl*, iii. 2. [3] *Ibid*, iii. 3.

This is certainly no improvement on the other method. It will, in fact, require more operations to complete the desired transformation. The first method is very general and is equally available for all cases of transformation of a rectangle into a square.

To transform a square into a rectangle.

Baudhāyana gives the following rule for transforming a square into a rectangle :

" If you wish to transform a square into a rectangle, divide it by the diagonal. Divide again one part into two, and add them suitably so as to fit the two sides (of the other half)."[1]

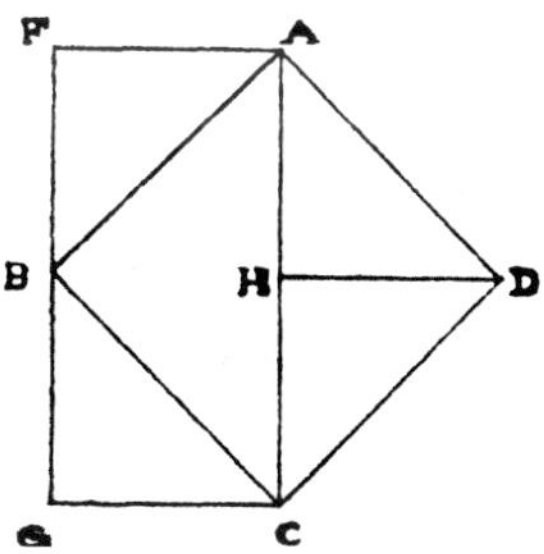

Fig. 46

The same method is also taught by Kātyāyana.[2] This method is much circumscribed inasmuch as it transforms the given square into a rectangle in which the length is double the breadth and is itself equal to the diagonal of the given square.

To transform a square into a rectangle which shall have a given side.

To transform a square into a rectangle which shall have a given side, Āpastamba gives the following rule :

" If you wish to transform a square into a rectangle,

[1] *BŚl*, i. 52. [2] *KŚl*, iii. 4.

(cut off from it a rectangular segment) by making a side as long as you wish (a side of the transformed rectangle to be). What remains in excess, should be added (to the former) as suitably as to fit."[1]

A similar rule is given also by Baudhāyana:

"Or else if the square is to be transformed into (a rectangle) of this (*i.e.*, one specified) side, cut off (from the square) a segment by that side. What remains in excess should be added along the other side."[2]

The latter portion of both the rules is obscure inasmuch as the operations to be employed have not been explained fully. They were doubtless handed down by oral tradition.

Thibaut,[3] followed by Bürk,[4] thinks the method implied to be this: Let the side of a given square be 7 units long. It is to be transformed into a rectangle whose one side will be, say, 5 units long. Cut off from the given square a rectangle of 5 by 7 units. There will then remain in excess a rectangle of 7 by 2 units. From this cut off a rectangle of 5 by 2 units and add it properly to the other portion. Then remains a square of 2 by 2 units. Change this into a rectangle of 5 by 4/5 units and place it by the side of the previous rectangle so as to fit. Thus we have finally a rectangle of 5 by 49/5 (=7+2+4/5) units.

This explanation is doubtless wrong. For at the final stage it begs the original problem itself. If the *Sulba* writers had really intended a simple arithmetical operation

[1] *ĀpŚl*, iii. 1.

[2] *BŚl*, i. 53. Thibaut's reading of the beginning of this sūtra is incorrect. It should be अपि वैतस्मिश्चतुरस्र' not अपि चैतस्मिंश्चतुरस्र' The former reading is given by Caland in his edition of the *Baudhāyana Śrauta Sūtra* and also appears in my copies of the *Baudhāyana Śulba*.

[3] *Śulba Sūtra*, p. 20.

[4] *ZDMG*, LVI, p. 334.

to be followed at the final stage of their method, as is supposed by Thibaut and Bürk, they could have, and very likely would have, directed to adopt it at the preliminary stage, without taking recourse to any kind of geometrical operations at all. The method of Āpastamba and Baudhāyana was of course geometrical, not arithmetical.[1]

According to the commentators Sundararāja and Dvārakanātha Yajvā, the method implied is similar to this:

"Produce the northern and southern sides towards the east as much as you wish (a side of the transformed rectangle to be. Complete the rectangle and) draw the diagonal passing through its north-eastern corner. (Find the point) where it cuts the transverse side of the (given) square lying inside that rectangle. Leave off the portion of that side lying to the north of that point and make its southern portion the breadth of a rectangle. That will be the rectangle (required)."[2]

[1] It may be noted that Thibaut has discarded on a different occasion certain arithmetical interpretation of a rule of Baudhāyana by Dvārakanātha Yajvā with the remark, "The commentary instead of showing how the desired end could be attained by making use of the geometrical constructions taught in the paribhāshā sūtras, employs arithmetical calculation; but this was of course not the method of the sūtrakâra." (*Paṇḍit* O. S., X, p. 73.)

Similarly Bürk remarks about certain explanation of the commentator Sundararāja: "This explanation of *jñeya* is incorrect...because the Śulbasūtra deals with geometrical construction, and not with 'numerical calculation.' In commenting *jñeyā* from the standpoint of arithmetic, the commentator falls into a similar error as disclosed by Thibaut in another case (*JASB*, XLIV, 272)." (*ZDMG*, LVI, p. 329.)

[2] "यावदिष्टं पार्श्वमान्यौ प्राच्यौ वर्द्धयित्वा उत्तरपूर्व्वां कर्णरज्जुमायच्छेत्। सा दीर्घचतुरस्रमध्यस्थायां समचतुरस्रतिर्य्यङ्मान्यां यत्र निपतति तत उत्तरं छित्वा दक्षिणांशं तिर्य्यङ्मानीं कुर्य्यात्। तद्दीर्घचतुरस्रं भवति।"

Let $ABCD$ be the given square and M the given length which is greater than a side of the square.

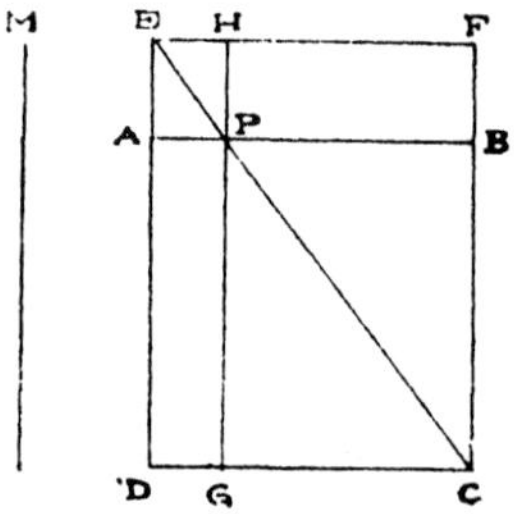

Fig. 47

Produce DA and CB to E and F respectively so that $DE=CF=M$. Join EF and complete the rectangle $EFCD$. Draw the diagonal EC cutting AB at P. Then PB will be the breadth of the transformed rectangle. Through P draw the straight line HPG parallel to ED or FC. Then $HFCG$ is the rectangle which is equivalent to the square $ABCD$ and whose side CF is equal to the given length M. For

$$\triangle EFC=\triangle EDC,$$
$$\triangle EHP=\triangle EAP,$$
$$\triangle PBC=\triangle PGC.$$

$\therefore$ parallelogram $HFBP$=parallelogram $APGD$.

Hence parallelogram $HFCG$=square $ABCD$.

Q. E. D.

Bürk was led to suspect whether the method explained by the commentators was indeed in view of the Śulbakāras and to discard it ultimately for the simple reason "that this method is so scientific." Thibaut has not directly assigned any reason for his rejection of the interpretation of the commentators. He seems, however, to have been led by an observation of Dvārakanātha Yajvā *anyacca prakāraḥ* ("Also another method") which just precedes his delivery of the method. But the preliminary remark of Sundararāja is *ayamatra prakāraḥ* ["This is the method (taught) here]." The rule has been formulated by the two commentators in identical words. So one has doubtless copied from the other. Difference in the preliminary remark may be explained thus: Sundararāja explains the method just after the above rule of Āpastamba which is its proper place. But Dvārakanātha Yajvā gives it under the imperfect method of Baudhāyana for the transformation of a square into a rectangle which precedes his perfect rule stated above. So it was very natural for him to remark that it is "another method."

Why Dvārakanātha Yajvā explains the method at a place other than what is proper for it seems very probably to be this. The process as defined in the rules of Baudhāyana and Āpastamba does not begin, as has been already pointed out by Bürk, with the construction of a greater rectangle (*EFCD* in Fig. 47) as required in the above explanations of the commentators. But it begins very clearly to cut off from the given square a smaller rectangle (*ABFE* in Fig. 48). Or in other words the explanation of the commentators has in view the case in which the given length is greater than a side of the given square, whereas the rule of the Śulbakāras has in view the case in which the given length is smaller than a side of the given square. So Dvārakanātha Yajvā was not wrong in calling the method described by him to be a different one. This case was treated under the unsatisfactory rule of Baudhāyana undoubtedly as a sort of modification and improvement of it. The case of the Śulbakāra has been explained by him in its proper place under the succeeding rule.

The geometrical process in the case that was in view of the Śulbakāras is this: From the sides *AD* and *BC* of the given square $ABCD$ cut off AE and BF respectively,[1]

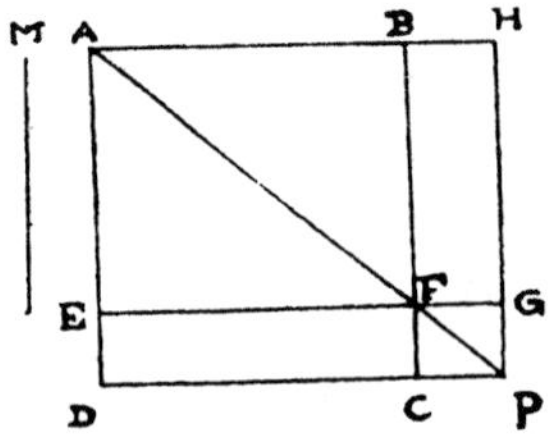

Fig. 48

making each equal to the given length M which is smaller than a side of the given square. Join AF and produce it to meet DC produced at P. Complete the rectangle $ADPH$. Produce EF to meet HP at G. Then $AHGE$ is the rectangle which is equivalent to the square $ABCD$ and which has a side AE equal to the given length M. The proof is similar to that given before.

On one occasion Baudhāyana makes a rectangle equivalent to three given squares, one side of the rectangle being half of a side of a square.[2] This case is a very simple one indeed, but it shows that he knew other methods of transforming a square into a rectangle than that indicated in his imperfect rule.

To transform a square or a rectangle into an isosceles trapezium which shall have a given face.

Baudhāyana gives the following rule for the transformation of a square or a rectangle into an isosceles trapezium, or for what they call "shortening of a square or a rectangle on one side."

[1] We cut off lengths equal to M from the sides AD and BC of the given square but not from the sides AB and CD, because the rule clearly directs that the given length M should be made the *pārśvamānī* of the rectangle cut off.

[2] *BŚl*, iii. 255.

"If you wish to make a square or a rectangle shorter on one side, (cut off a rectangular portion) by making the shorter length a side. Divide the remainder by the diagonal and place (the two portions) on either sides (of the portion cut off) after inverting."[1]

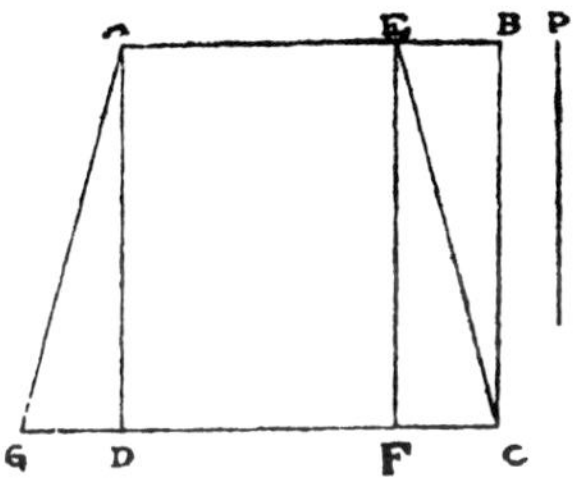

Fig. 49

Let $ABCD$ be a given square and P a given line which is shorter than AB. From AB and DC cut off AE and DF respectively making each equal to P. Join EF and EC. Take the triangle CBE and place it after inverting in the position ADG. Then $AECG$ is the isosceles trapezium which is equal to the given square $ABCD$ and whose face AE is equal to the given length P.

We find a nearly similar rule in the *Satapatha Brāhmaṇa.*[2] Let $ABCD$ be a rectangle. Take $AE = FB =$

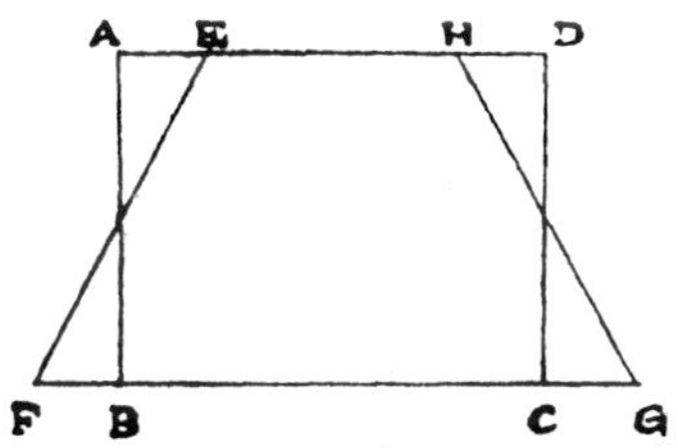

Fig. 50

$DH = CG$. Then it is said that the trapezium $EFGH$ is

[1] *BŚl*, i. 55.

[2] *ŚBr*, x. 2. 1. 4.

exactly equal to the rectangle *ABCD*. This method reappears in the *Āpastamba Śulba*.[1]

To transform a square or a rectangle into a triangle.

" If you wish to transform a square or a rectangle into a triangle, construct a square whose area will be twice as much as the area of the figure (to be transformed). Fix a pole at the middle of its eastern side. Having fastened at it two ties (of two cords), stretch the cords towards the two western corners. Cut off the portions lying beyond these cords." [2]

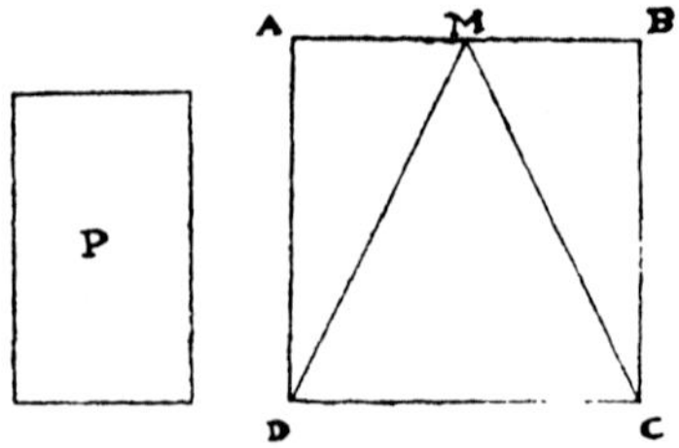

Fig. 51

Let *P* be the given rectangular figure to be transformed. Draw the square *ABCD* so that its area will be twice that of *P*. Let *M* be the middle point of *AB*. Join *MD* and *MC*. Then the triangle *MCD* is equal to the rectangular figure *P*. For each is equal to the half of the square *ABCD*.

To transform an isosceles triangle into a square.

" If you wish to transform an isosceles triangle into a square, cut off its northern half by the middle line ; then place it on the opposite side after inverting, By the method of constructing a square equal to a rectangle,

[1] *ĀpŚl*, xv. 9f.
[2] *BŚl*, i. 56.

construct the square. This is the method of construction." [1]

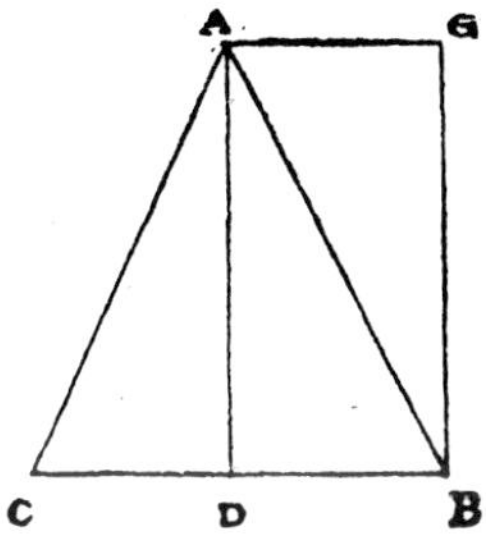

Fig. 52

Let ABC be an isosceles triangle. Draw the median AD. Take the half ADC and place it after inversion in the position BGA. Then the rectangle $AGBD$ is equal to the isosceles triangle ABC. Now transform the rectangle $AGBD$ into a square by the method given before.

To transform a square or a rectangle into a rhombus.

" If you wish to transform a square or a rectangle into a rhombus, construct a rectangle which shall have an area twice as much as the area (of the figure to be transformed). Fix a pole at the middle of its eastern side. Having fastened at it two ties (of two cords), stretch the cords towards middle (points) of the northern and southern sides (of the rectangle). Cut off the portions lying beyond (these cords). Thereby is also explained the construction of the other triangle." [2]

Let P be a rectangular figure. Draw the rectangle $ABCD$ so that its area will be double that of P. Let G, H, E, F be the middle points of the sides AB, BC, CD,

[1] *KŚl*, iv. 5.
[2] *BŚl*, i. 57.

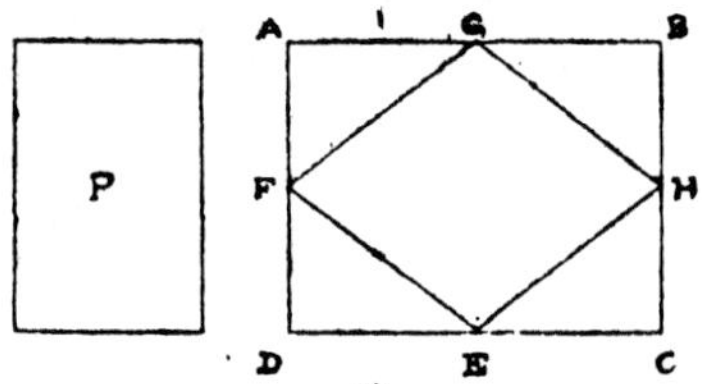

Fig. 58

DA respectively. Join *GH*, *HE*, *EF*, *FG*. Then the rhombus *GHEF* is equal to the rectangular figure *P*.

This method is taught by Baudhāyana, Āpastamba[1] and Kātyāyana.[2]

To transform a rhombus into a square.

Kātyāyana observes:

" If it be the case of (transformation into a square of) a rhombus, bisect it by its transverse middle line. Then construct as before."

That is, the rhombus is first divided into two isosceles triangles by drawing a diagonal. The triangles are then transformed into squares by the method given before. Finally, the two squares are combined into one square.

1 *ĀpŚl*, xii. 9.
2 *KŚl*, iv. 4.
3 *Ibid*, iv. 6.

CHAPTER VIII

Areas and Volumes

The unit of area is defined by Āpastamba thus:

"By means of a measure is produced a measure."[1] That is, the unit of surface measure or area is the area of a square on a side of unit length.

In the *Śulba* we find express rules for the mensuration of squares and isosceles trapeziums only. It is certain that mensuration of certain other elementary figures, such as triangles and rectangles, was also known in that time. The area of the circle was only roughly approximate. Other kinds of rectilinear figures, particularly the Fire-altars of various shapes enumerated before, were used to be mensurated by breaking them up into elementary triangles, squares and rectangles.

The number of square units in the area of a square is obtained by multiplying the number of linear units in a side by itself.

Āpastamba and Kātyāyana state:

"As many units of a measure as are in a cord, so many rows (or series) of squares (of that measure) there will be in a square on that cord as a side."[2]

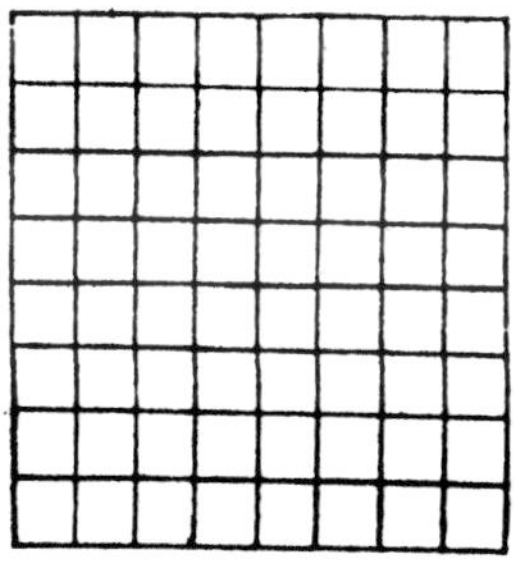

Fig. 54

1 *ĀpŚl*, iii. 4.

2 *ĀpŚl*, iii. 7; *KŚl*, iii. 9.

This result was of course generalised, so that the area of any square was used to be calculated by multiplying its side by itself.

The number of square units in the area of a rectangle is obtained by multiplying together the numbers of linear units in the length and breadth of the rectangle.

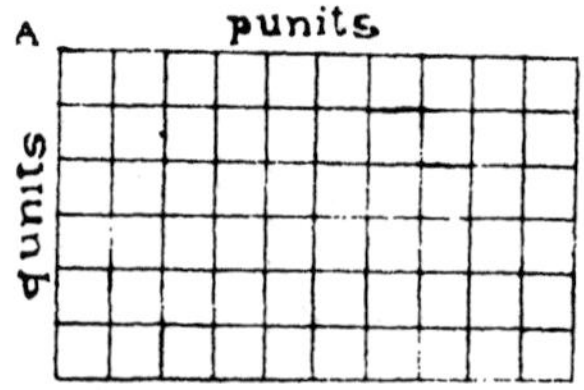

Fig. 55

Let $ABCD$ be a rectangle whose length AB contains p units and whose breadth AD has q units. On dividing AB and AD and drawing parallel lines, it is found that the rectangle $ABCD$ is divided into pq unit squares. Therefore its area is pq units of area.

It should be noted that we do not find in the *Śulba* any explicit rule for the mensuration of a rectangle. But there is no doubt that it was used to be done in the above way. Similarly from the method of constructing the *ubhayī* bricks (Fig. 7) which are of the shape of a scalene triangle, and of the parallelograms (Figs. 32-5 ; *cf.* also Fig. 14), it is clear that the following formulae also were known :

Area of a triangle $= \frac{1}{2}$ (base) × (altitude),
Area of a parallelogram = (base) × (altitude).

The area of a trapezium.

How to find the area of a trapezium has been *demonstrated* by Āpastamba in the course of determination of the area of the *Mahāvedi* which is of the shape of an isosceles trapezium whose altitude, face and base are respectively 36, 24 and 30 padas (or prakramas). He says:

" The *Mahāvedi* measures (in area) one-thousand less twenty-eight (square) padas. Draw a straight line from the south-eastern corner (of the *vedi*) to a point 12 padas towards the south-western corner. Place the portion thus cut off on the other (*i.e.*, northern) side of the *vedi* after inverting it. It (the *Mahāvedi*) will then become a rectangle. After that construction the area will be apparent." [1]

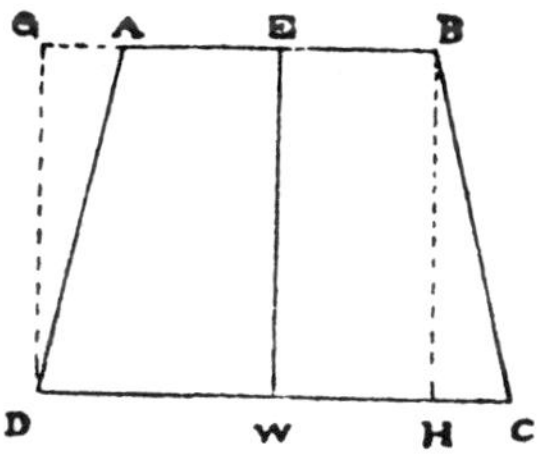

Fig. 5·6

Let $ABCD$ be the isosceles trapezium. Cut off $WH = EB$. Join BH. Place the triangle BHC after inversion in the position DGA. Then, the rule says, as is also obvious,

The area of the trapezium $ABCD$

$$= \text{the area of the rectangle } GBHD,$$
$$= DH \times BH,$$
$$= \tfrac{1}{2}(AB + DC) \times BH,$$
$$= \tfrac{1}{2}(\text{face} + \text{base}) \times \text{altitude}.$$

This result also follows at once from the method indicated in the *Śatapatha Brāhmaṇa* and by Baudhāyana for the transformation of a square or a rectangle into an isosceles trapezium.[2]

1 *ĀpŚl*, v. 7.

2 *Vide supra*, p. 90f.

Figures with a given area.

For the construction of an altar of proper size and shape, as prescribed by the Holy Scriptures, it is sometimes necessary to describe a rectilinear figure having a given area. For instance, it is prescribed that the *vedi* of the *Pitṛyajña* must be a square of an area equal to one-ninth of that of the *Mahāvedi.*[1] Now the area of the *Mahāvedi* is given to be 972 square padas. So the problem becomes to construct a square having an area of 108 square padas. Again it is said that the *Sautrāmaṇikī-vedi* is of the form of an isosceles trapezium having an area equal to the third part of the *Mahāvedi*, that is, to 324 square padas. According to Āpastamba and Kātyāyana, this isosceles trapezium must be similar to the shape of the *Mahāvedi.* This case will be treated separately under "Similar figures." But Baudhāyana seems to have left the shape of the trapezium unrestricted. Hence there arises the problem: to find an isosceles trapezium having an area of 324 square padas. According to the tradition of some schools, it is necessary to construct square-shaped altars of area varying from $1\frac{1}{2}$ to $6\frac{1}{2}$ square puruṣas.[2]

To construct a square having an area of 108 square padas.

Baudhāyana gives the following method:

" One-third producer (*i.e.*, the side of a square whose area is one-third the area) of the square made with the third part of the *Mahāvedi* is that (the side of the *Paitṛki-vedi*). Its area is the ninth part of the area (of the *Mahā-vedi*)."[3]

$108 = 324/3 = 18^2/3$. So the required square will be one-third of a square on a side 18 padas long. The

[1] *BŚl*, i. 81.

[2] *BŚl*, ii. 14; *ĀpŚl*, viii. 3.

[3] *BŚl*, i. 82; compare *Paṇḍit*, X (O. S.). p. 46f

method for this latter construction has been taught before. Let AB be a straight line 18 padas long. Divide it into three equal parts. Let AC be one such part. Describe the square $ACDE$. Join AD. Describe a circle with

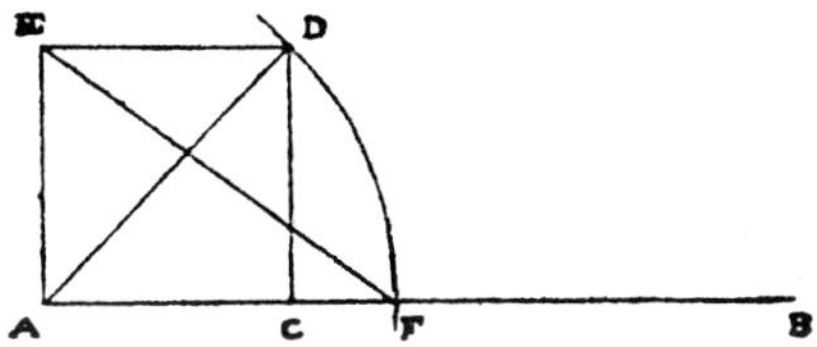

Fig. 57

centre A and radius AD cutting AB at F. Join EF. Then EF is a side of the square having an area of 108 square padas. For

$$\begin{aligned} EF^2 &= EA^2 + AF^2, \\ &= EA^2 + AD^2, \\ &= 3AC^2, \\ &= \tfrac{1}{3}AB^2. \end{aligned}$$

$$\therefore \quad EF^2 = 108 \text{ square padas.}$$

This construction is the same as to find a square three times a square of 36 sqaure padas. For $108 = 3 \times 6^2$. If 108 would have been a square number or expressible as the sum of the two square numbers, the solution would have of course been very easy.

To construct an isosceles trapezium having an area of 324 square padas.

Baudhāyana says:

" If a square be formed with the third part of the *Mahāvedi*, its sides will be each 18 padas long. Then by making it longer on one side and shorter on the other, the sides should be determined optionally as necessary." [1]

[1] *BŚl*, i. 86-7.

The rule does not explain clearly how the sides of the square formed with the third part of the *Mahāvedi*, are to be varied, so that the modified figure may assume the shape of an isosceles trapezium, but still retaining the same area. It seems, however, that the altitude is left unchanged, only the face and base being varied.

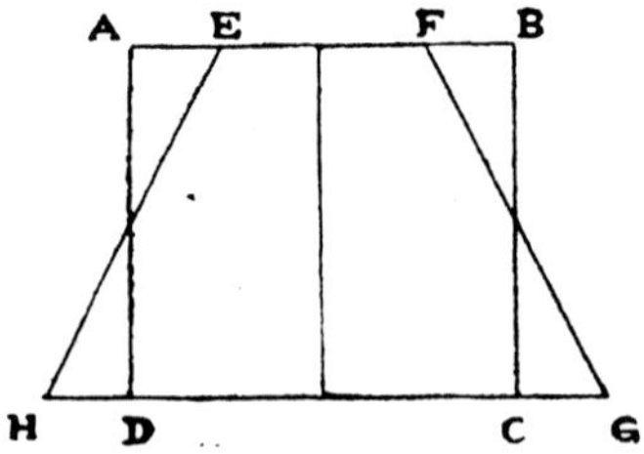

Fig. 58

Let $ABCD$ be the square whose side AB is 18 padas long. Suppose $EFGH$ to be the modified form. Also suppose $EF=18x$ and $HG=18y$. Since the area must remain the same, we must have

$$18\left(\frac{18x+18y}{2}\right)=324,$$

or

$$x+y=2.$$

Thus we can easily obtain any number of isosceles trapeziums having the same altitude and area.

We have explained before Baudhāyana's method for the transformation of a square (or a rectangle) into an isosceles trapezium which shall have a given face. That is different from the one supposed here. But this method has one advantage over that ; it does not disturb the east-west line and so its adoption, we think, to have been more probable. If the altitude also vary, we get an indeterminate equation of the form

$$z(x+y)=2$$

where $18z$ is the altitude of the resulting isosceles trapezium.

Volume of a Prism or Cylinder.

We have so far treated of the shape and size of the base of the various kinds of Fire-altars and have observed that the size in all cases has been prescribed to be the same, namely, $7\frac{1}{2}$ square puruṣas. Every Fire-altar, save and except the one of the shape of the cemetery (*Śmaśāna-cit*), has a perfectly horizontal surface whose height is prescribed to be one *jānu* (=32 aṅgulis) for the first construction, two *jānus* for the second construction, and so on. Again the number of layers of bricks at successive constructions increases as multiples of five ; and the shape and size of any layer of a particular Fire-altar at any construction is the same as those of its base. Hence all the Fire-altars are right prisms or circular cylinders. And the early Hindu geometers evidently knew the formula

$$\text{Volume of a prism or cylinder} = (\text{base}) \times (\text{height}).$$

Approximate Volume of the Frustum of a Pyramid.

The cemetery-shaped Fire-altar is in fact a frustum of a pyramid. Its base is an isosceles trapezium whose dimensions are stated by Baudhāyana to be as follows :

"He should construct the *Śmaśāna-cit* ("the Fire-altar of the shape of the cemetery") who desires : 'May I gain prosperity in the *Pitṛloka* (the world of the fathers) :'[1] it has been (taught). Six puruṣas are the length of the east-to-west line, three the length of the eastern side and two the length of the western side. This is the body (of that Fire-altar)."[2]

[1] The quotation is from *TS*, v. 4. 11.3.

[2] *BŚr*, xvii. 30.

It should be remarked that the unit puruṣa in this passage is not the ordinary puruṣa of 120 aṅgulis, but a reduced unit whose length is equal to a side of a square equivalent to one half of the ordinary square puruṣa.[1] So the area of the trapezium is equal to 15 reduced square puruṣas or 7½ ordinary square puruṣas.

The height of this Fire-altar has been specified by Baudhāyana thus:

"Its (*Śmaśāna-cit*) measure is, when neck-deep on the east, navel-deep on the west; when navel-deep on the east, knee-deep on the west; when knee-deep on the east, ankle-deep on the west; and when ankle-deep on the east, it is on a level with the ground on the west."[2]

Notwithstanding this difference in height on the two sides of the Fire-altar, its cubic content has to be kept intact. To effect that in practice the following device is generally adopted:

"Increase the (usual) vertical measure of the Fire-altar by its one-fifth. Then divide the total height into three parts, and make bricks with (the height equal to) the fourth, ninth or fourteenth part of two of these parts. Construct four, nine or fourteen layers with them. Divide the remaining part (having constructed it with one layer of bricks of the height of one-third the total height) by the diagonal (plane) inclined down towards the west and remove the half (*i.e.*, the upper portion)."[3]

It has been prescribed that the nth construction of the Fire-altar shall have a height of n jānus and comprise of $5n$ layers of bricks. Increasing the height by its one-fifth, we get $6n/5$ jānus. Two-thirds of them gives $4n/5$ jānus. Up to this height, the altar is constructed in $(5n-1)$

[1] *BŚl*, iii. 253-4; compare also Thibaut, *Śulrasūtra*, p. 40.

[2] *BŚr*, xvii. 30.

[3] *BŚl*, iii. 266-8..

layers, so that the height of each brick is equal to $(5n-1)$th part of $4n/5$. One-third of the increased altitude is $2n/5$ jānus. Next layer of the Fire-altar is constructed with bricks of height $2n/5$ jānus. Then the upper portion of this layer is cut off by the diagonal plane as directed. Hence the altitude of the altar is now $6n/5$ jānus on the east, $4n/5$ jānus on the west ; so that its average altitude is $(6n/5+4n/5)/2$ or n jānus. This device will be easily recognised to be based on the following approximate formula for calculating the volume of the frustum of a pyramid: If (a, b) be the length and breadth of the rectangular base of the solid, (a', b') the corresponding sides of the face parallel to it and h the height, then

$$\text{Volume of the frustum} = \left(\frac{a+a'}{2}\right)\left(\frac{b+b'}{2}\right)h.$$

CHAPTER IX

The Theorem of the Square of the Diagonal

There is one very important proposition which, together with its converse, looms much more considerably than anything else through the entire geometry of the *Śulba*. Baudhāyana has enunciated it thus:

" The diagonal of a rectangle produces both (areas) which its length and breadth produce separately." [1]

That is: the square described on the diagonal of a rectangle has an area equal to the sum of the areas of the squares described on its two sides. The proposition is defined in almost identical terms also by Āpastamba [2] and Kātyāyana. [3]

This theorem is now universally associated with the name of the Greek Pythagoras (*c.* 540 B.C.), though " no really trustworthy evidence exists that it was actually discovered by him." [4] To denote it as shortly and clearly as possible, Hankel suggested for it the name " the Theorem of the Square of the Hypotenuse." It would be more in keeping with the form and spirit of the early Hindu geometrical terminology to alter this slightly to "the Theorem

[1] " दीर्घचतुरश्रस्याक्ष्णयारज्जुः पार्श्वमानी तिर्यङ्मानी च यत्पृथग्भूते कुरुतस्तदुभयं करोति ।"—*BŚl*, i. 48.

Compare *BŚr*, x. 19 ; xix. 1 for applications of the theorem.

[2] " दीर्घस्याऽक्ष्णयारज्जुः पार्श्वमानी तिर्यङ्मानी च यत् पृथग्भूते कुरुतस्तदुभयं करोति ।"—*ĀpŚl*, i. 4.

[3] " दीर्घचतुरश्रस्याक्ष्णयारज्जुस्तिर्यङ्मानी पार्श्वमानी च यत् पृथग्भूते कुरुतस्तदुभयं करोतीति क्षेत्रज्ञानम् ।"—*KŚl*, ii. 11.

[4] Heath, *Greek Math.*, I, p. 144f.

of the Square of the Diagonal." For the *Sulba* does not speak, as we do, of the right-angled triangle, but of the square and the rectangle. Hence the altered title is adopted here.

The employment of two different terms by compartment for what we denote by one, necessitated the ancient Sulbakāras to define the above proposition for the rectangle again with reference to the square:

"The diagonal of a square produces an area twice as much." [1]

That is: The area of the square described on the diagonal of a square is double the area of that square.

The order in which these two and other closely related propositions is mentioned by different writers is noteworthy. The oldest known Sulbakāra, Baudhāyana, states the second proposition before the first intervened by two other rules in the same connexion. The posterior writers like Āpastamba and Kātyāyana place the second just after the first. This change in the order in stating these propositions may not be entirely without any significance. It shows that by the time of Āpastamba and Kātyāyana the importance and generality of the theorem of the square of the diagonal of a rectangle was recognised fully. So they very naturally stated the corresponding theorem for the particular case of a square as a corollary to it. But from the point of view of the origin and growth of the theorem, Baudhāyana's arrangement is more natural.

Another proposition which has been most freely used in the *Sulba* for the construction of a right angle is this:

"If a triangle is such that the square on one side of it is equal to the sum of the squares on the two other sides, then the angle contained by these two sides is a right angle."

[1] *BSl*, i. 45; *ĀpSl*, i. 5; *KSl*, ii. 12. For applications of the theorem compare *BSr*, x. 19; xix. 1, xxvi. 10.

This converse proposition is not found explicitly defined by any Śulbakāra but its truth is tacitly assumed by all of them.

Did the ancient Hindus discover a proof of the theorem of the square of the diagonal? No conclusively satisfactory answer which will be beyond a shadow of doubt can be given to this question. For there is not found any mention, direct or indirect, of such a proof, even if it had existed anywhere in the early literature of the Hindus. So what we shall say on this point will be more or less conjectural, based of course on other matters having positive bearing on the point in question. It may be pointed out that the state of affairs is not much better elsewhere. It has been already stated that though the proposition is now universally associated with the name of the Greek Pythagoras, "no really trustworthy evidence exists that it was actually discovered by him." The tradition which attributes the theorem to him began five centuries after Pythagoras and was based upon a vague statement which did not specify this or any other great geometrical discovery as due to him. This led some eminent scholars like Hankel [1] and Junge [2] even to deny to Pythagoras the discovery of the proposition of the theorem of the square of the diagonal. Again the method which is supposed by the modern believers of the Pythagoras hypothesis to have been presumably followed by him to prove the theorem is purely conjectural.[3] So in the same way we proceed to discuss how the theorem was proved by the ancient Hindus. It may be noted

[1] H. Hankel, *Zur Geschichte der Math. in Alterthum und Mittealter*, Leipzig, 1874, p. 97.

[2] G. Junge, "Wann haben die Grieschen das Irrationale entdeckt?" *Novae Symbolae Joachimicae*, Halle, 1907, pp. 221-264; quoted by Heath (*Euclid*, I, p. 351).

[3] Heath, *Euclid*, I, pp. 352 ff.

that Bürk and Hankel are definitely of opinion that the Hindus had a general geometrical proof. So also was, and more pronouncedly, the eminent German philosopher Schopenhauer.

The rule of Baudhāyana immediately following that containing the enunciation of the general theorem of the square of the diagonal runs thus:

"This (*i.e.*, the truth of the theorem) is perceived in the rectangles with sides of three and four (units), twelve and five, fifteen and eight, seven and twenty-four, twelve and thirty-five, fifteen and thirty-six (units)." [1]

The Sanskrit terms are *trikacatuṣkayoḥ*, etc. They literally mean, "the rectangle whose sides have three (units) and four (units)," etc.

From this one might surmise that the ancient Hindus were aware only of the arithmetical character of the theorem and then by an imperfect generalisation applied it to rational rectangles. But such a surmise will be too hasty. For there cannot be absolutely any doubt about the fact that the Hindus fully recognised the most general geometrical character of the theorem and employed its truth universally. Indeed, we find instances of its application to cases of rectangles whose sides cannot be represented by rational quantities. For instance, for the construction of the *Sautrāmaṇīkī-vedi*, application is made of the right-angled triangle $(15/\sqrt{3}, 36/\sqrt{3}, 39/\sqrt{3})$ or $(5\sqrt{3}, 12\sqrt{3}, 13\sqrt{3})$ and for the *Aśvamedhikī-vedi* of the right-angled triangle $(15\sqrt{2}, 36\sqrt{2}, 39\sqrt{2})$.

The propositions about the combination and transformation of areas which we have noticed before may also be cited in this connexion. Perfectly geometrical character of them cannot be questioned.

[1] *BŚl*, i. 49.

Above all there is the remark of Kātyāyana at the end of his enunciation of the general theorem of the square of the diagonal of a rectangle: *iti kṣetrajñānaṃ* or "this is the knowledge of (plane) figures." Thibaut renders it as "this is the knowledge (requisite) for (the measurement of) areas." This is evidently inaccurate. For in the *Śulba-sūtra*, the area is technically called *bhūmi*, not *kṣetra* which denotes "figures." These prove conclusively that the universal geometrical character of the theorem was fully recognised.

On the other hand the above rule of Baudhāyana suggests that the truth of the theorem of the square of the diagonal was perceived and proved in the case of rational rectangles first; and it was then generalised and found to be true universally. This is perfectly natural. In support of this hypothesis may be cited the rule of Āpastamba and Kātyāyana for the calculation of the area of a square:

"As many units of a measure as are in a cord so many rows (or series) of squares (of that measure) there will be in a square on that cord as a side." [1]

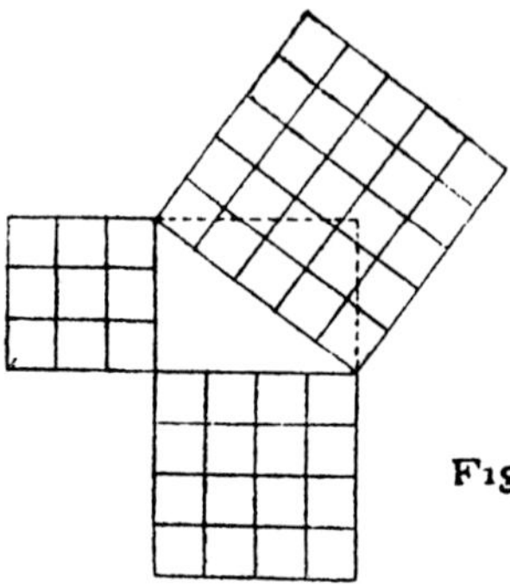

Fig. 59

So that by drawing the squares on the sides and diagonal of a rational rectangle and dividing them into

[1] *ĀpŚl*, iii. 7; *KŚl*, iii. 9. *Vide supra* p. 95.

elementary squares it will be easily found by calculation that the square on the diagonal is equal to the sum of the squares on the sides.

This hypothesis as regards the proof of the theorem presupposes a knowledge of the rational rectangles. How the ancient Ḥindus discovered such rectangles we shall discuss in the next chapter.

The rules of Baudhāyana just preceding the one containing the proposition of the theorem of the square of the diagonal of a rectangle are these:

"The diagonal of a square produces an area twice as much.

"(Take a rectangle whose) breadth is (equal to) the measure (of the side of a square) and length (equal to) its *dvikaraṇī*; its diagonal will be *trikaraṇī* ('three-fold-producer' of the square).

"Thereby is explained the *tṛtīya-karaṇī* ('the generator of the third part' of the square); it is the ninth part of the area."[1]

If this arrangement of the propositions can be supposed to give any clue as to the discovery of the theorem of the square of the diagonal, then it will have to be said that the theorem for a square was discovered first. In that case the hypothesis about the discovery of a proof of the theorem must be a little different from the one suggested above.

Now one of the oldest Hindu Fire-altars is the *Catura-sra-śyenacit*. The oldest method for its construction does not presuppose a knowledge of the theorem in question. This method had been expressly taught by Āpastamba. It was undoubtedly known to Baudhāyana who hints it very briefly and gives in fact what is rather an improved form of it. Bürk[2] surmises that the proof of the theorem was

[1] *BŚl*, i. 45-7; see also *BŚr*, xix. 1.

[2] Bürk, *ZDMG*, LV, pp. 556 f.

discovered just in the figure of the *Caturasra-śyenacit*. The square *ACFE* on the diagonal *AC* of the square *ABCD* of the four squares forming the *ātman* (or "body")

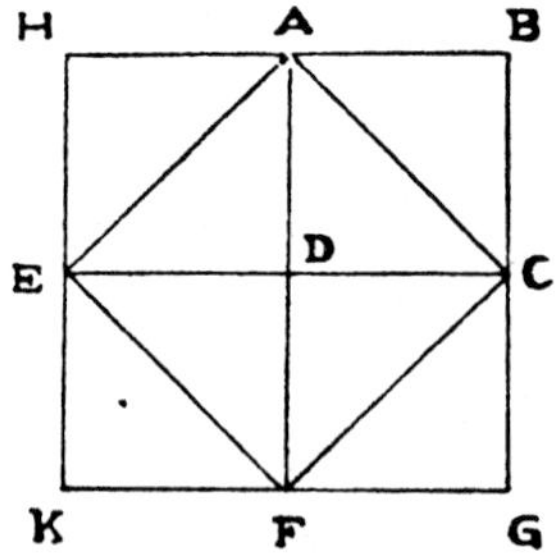

Fig. 60

of this altar is obviously equal to the square *ADEH* on the side *AD* and the square *DCGF* on the side *DC*. Bürk has further confirmed his hypothesis by a reference to Baudhāyana's imperfect rule (taught also by Kātyāyana) for the transformation of a square into a rectangle.

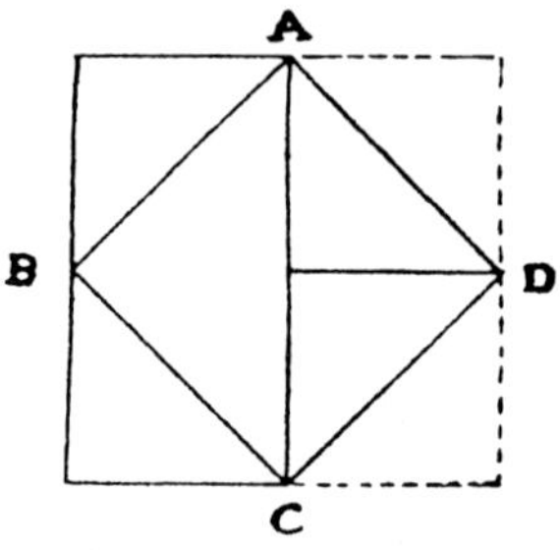

Fig. 61

This hypothesis about the discovery of the proof of the theorem of the square of the diagonal of a square is endorsed also by Heath.[1] It is certainly more likely than the one suggested by Cantor [2] and Allman [3] about Pythagoras' proof of the theorem. (Fig. 62.)

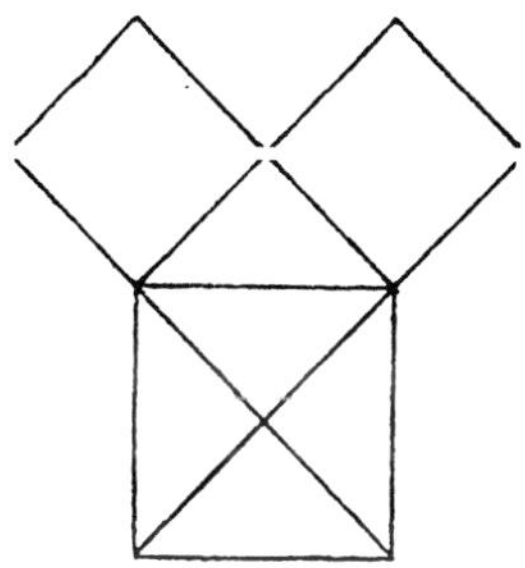

Fig. 62

Thibaut says: " The authors of the sūtras do not give us any hint as to the way in which they found their proposition regarding the diagonal of a square; but we suppose that they, too, were observant of the fact that the square of the diagonal is divided by its own diagonals into four triangles, one of which is equal to half the first square [Fig. 63]. This is at the same time an immediately convincing proof of the Pythagorean proposition as far as squares or equilateral rectangular triangles are concerned." [4]

[1] T. L. Heath, *The Thirteen Books of Euclid's Elements*, in 3 volumes, Cambridge, 1908, Vol. I, p. 352.

It may be noted that this proof of the particular case of the theorem of the square of the diagonal was adduced with the figure, as that of its general case, by Al-khowârizmî (*c.* 825). (F. Rosen, *The Algebra of Mohammed Ben Musa*, London, 1831, pp. 74f.)

[2] M. Cantor, *Vorlesungen über Geschichte der Mathematik*, 3rd ed., Bd. I, p. 185.

[3] J. C. Allman, *Greek Geometry from Thales to Euclid*, Dublin, 1889, p. 29.

[4] Thibaut, *Sulbasūtras*, p. 8.

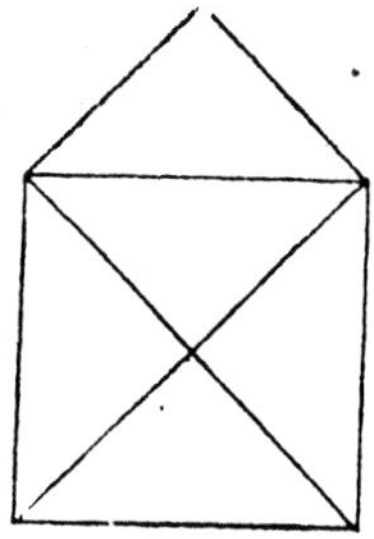

Fig. 63

Bürk thinks that this supposition is less probable, because it has no connecting link with the constructions ordinarily met with in the *Sulba*, also because it does not explain what might have induced the Sulbakāras, after they had drawn one square, to construct a new square on the diagonal of the same.

Such construction is not really as unnatural for the Sulbakāras as is supposed by Bürk. It is indeed clearly in evidence in the pattern of the first layer of bricks in the first kind of construction of the *Vakrapakṣa-śyenacit* as described by Baudhāyana.[1] To draw the attention directly, the relevant portions of it are here marked with bold lines [2] (Fig. 64). Further Baudhāyana teaches us to construct a square (brick) with half the diagonal of another square.[3] This leads to a construction of the kind supposed by Thibaut.

However, similar objections may also be raised, partly at least, against the supposition of Bürk. What might have led to the drawing of the diagonals of the four squares forming the body of the *Caturasra-śyenacit* ? They are not required as a matter of course.

[1] *BŚl*, iii. 62-104.

[2] *BŚl*, iii. 87-104. Compare *Paṇḍit*, Vol. X (O. S.), p. 211.

[3] *BŚl*, iii. 289.

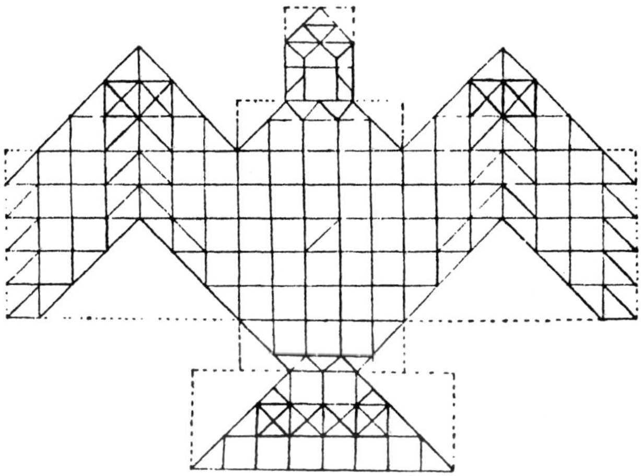

Fig. 64

Vakrapakṣa-syenacit

First layer of construction (*after Baudhāyana*)

An equally convincing hypothesis, which is very nearly alike to that of Bürk but free from all those defects, is suggested by the method of construction of the *Paitṛkī-vedi*. According to one tradition about that *vedi*, it is a square of one square puruṣa in area whose corners are turned towards the cardinal directions.[1] For its construction Kātyāyana indicates the following method:

"For the *Paitṛkī(-vedi)*, construct a square whose area is two square puruṣas; fix poles at the middle of its sides. (The figure formed by lines joining these poles will be the *vedi* required.) This is the method of construction."[2]

Mention of that tradition about the *Paitṛkī-vedi* is made also by Baudhāyana.[3] In fact, the construction of

[1] *KŚr*, xxi. 3. 28. [2] *KŚl*, ii. 6. [3] *BŚl*, i. 83-4.

a square with its corners pointed towards the cardinal directions is mentioned as early as the *Śatapatha Brāhmaṇa.*[1] But how to do it has not been indicated there. This is to be accounted by the fact that the method, which is undoubtedly the same as that described by Kātyāyana, was too well known. In fact it seems to have been the usual practice in all such cases. Compare Baudhāyana's imperfect method for the transformation of a square into a rectangle where also it is necessary in the beginning to construct a square whose corners are turned towards the cardinal directions. The common methods for the transformation of a square into a triangle and a double triangle or rhombus may also be referred to.

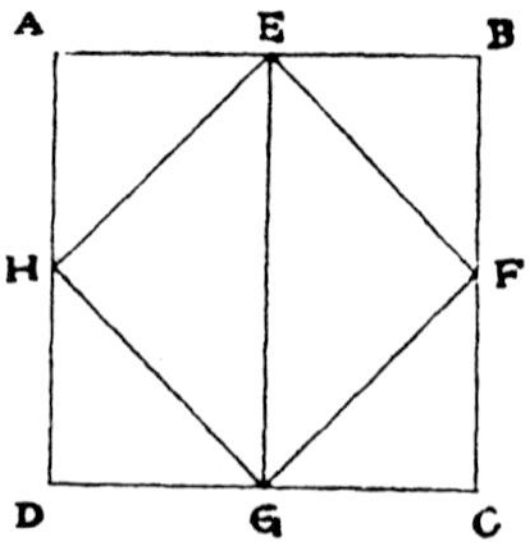

Fig. 65

It is thus learnt that the figure[2] $EFGH$ obtained by joining the middle points of the sides of a square $ABCD$ was known to be square in shape and half the original square in area. Now the original square $ABCD$ was naturally recognised as equivalent to the square on EG which is the east-west line. For as has been described

[1] *ŚBr*, xiii. 8. 1. 5.

[2] This figure was happily conceived also by Hankel as the possible source of the discovery of the theorem. But he was not aware of the special aspect it had in the geometry of the *Sulba.*

before the usual practice of the Hindus for the construction of a square (or indeed any other regular figure) of given sides was to construct it in such a way as to make it lie symmetrically on the east-west line. This EG is again the diagonal of the newly formed square $EFGH$. So this figure leads in a very simple and vivid way to the discovery and proof of the theorem of the square of the diagonal of a square. If we join HF, we at once obtain the construction forming the basis of Bürk's hypothesis.

How the ancient Hindus proceeded next to find a general proof is well hinted by the following two propositions of Kātyāyana preceding that of the general theorem of the square of the diagonal of a rectangle:

"(Take a rectangle whose) breadth is one pada and length three padas; its diagonal is the 'ten-fold-generator' (*i.e.*, it generates a square ten times as large as a square of one pada).

"(Take a rectangle whose) breadth is two padas and length six padas; its diagonal is the 'forty-fold-generator' (*i.e.*, it produces a square forty times as large as a square of one pada)." [1]

It is evident from Fig. 66 that the square $ABCD$ is equal to ten elementary squares, four forming the inner square $OPQR$ and the remaining six from the halves of

[1] *KŚl*, ii. 8-9. Compare *KŚr*, v. 3. 33, for an application of the first rule and xvii. 3.14 of the second.

In the manuscripts of the *Kātyāyana Śulba* known to me, there occurs a rule intervening between these two propositions and the proposition of the general theorem of the square of the diagonal:

"The measure of the *yuga* and the *śamyā* has been taught (before as it is found (in the holy scriptures)."—*KŚl*, ii. 10.

I think this rule has been placed here erroneously by a copyist. So there is a gradual development of the proposition of the theorem of the square of the diagonal from the particular cases to the most general one.

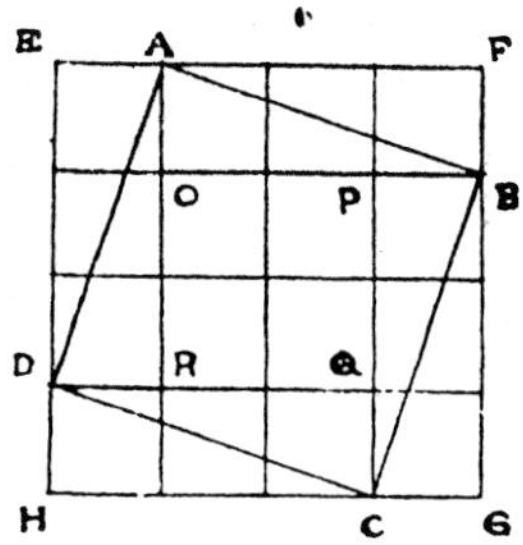

Fig. 66

the four rectangles surrounding it, *viz.*, *AFBO*, *BGCP*, *CHDQ*, *DEAR*, each of which contains three elementary squares. These can again be divided into two groups: one group consisting of nine elementary squares forming the square on the line *OB* and another group of a single elementary square on the side *OA*. Thus it is proved that

$$AB^2 = OA^2 + OB^2$$

If the side of each elementary square be one pada, we find a proof of Kātyāyana's first proposition, and if it measures two padas, a proof of his second proposition.

From these and similar instances of rectangles whose lengths and breadths can be represented by commensurable quantities, and in which the truth of the theorem is proved easily, it is not difficult to surmise how to proceed to find a general geometrical proof of it.

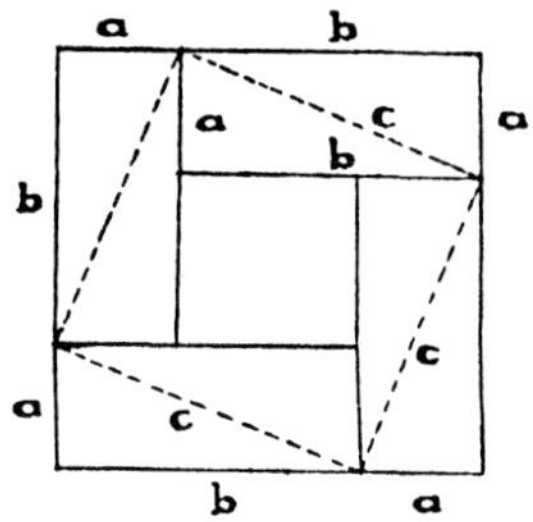

Fig. 67

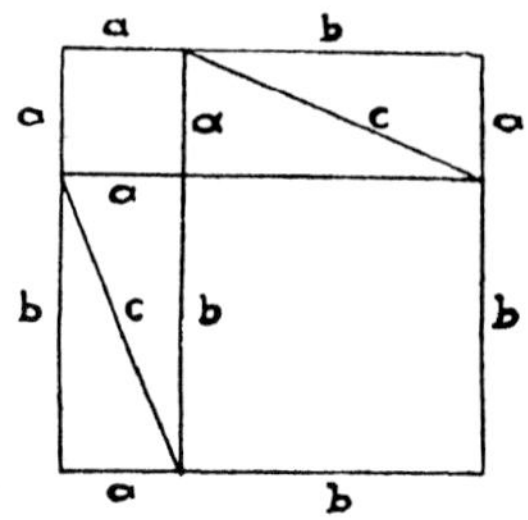

Fig. 68

According to this supposition we are to draw four rectangles equal to the given one each having as its diagonal a side of the square on the diagonal of the given rectangle. Then it follows obviously,

$$c^2 = 4(\tfrac{1}{2}ab) + (a-b)^2,$$

$$\text{or} \quad c^2 = a^2 + b^2.$$

We thus find from the *Śulba* how by successive stages the ancient Hindus developed, as is highly probable, a general proof of the theorem of the square of the diagonal. As confirmatory to this hypothesis, we may refer to the method of Āpastamba for the enlargement of a square. If it be required to construct a square whose side will exceed a side b of a given square by a, add, says Āpastamba,[1] on the two sides of the given square two rectangles whose lengths are equal to b and breadths to a; then add on the corner a square whose sides are equal to the increment a. Thus will be obtained a square with a side equal to $a+b$ (Fig. 68). A similar method is taught by Baudhāyana.[2] Now we can divide the added rectangles by their diagonals and place the four resulting triangles of sides a, b, c around another square of the same size as the enlarged square, in the manner shown in Fig. 67.

Nearly the same figure as Fig. 67 is formed by the constructions described in the *Śulba*, for the combination of two different squares (Fig. 68). The general truth of the theorem was very likely perceived from that figure.[3]

[1] *ĀpŚl*, iii. 9. See also p. 176 *infra*.

[2] *BŚl*, iii. 192-4.

[3] Compare C. Müller, "Die mathematik der Śulvasûtra," *Abhand. a. d. math. seminar d. Hamburgischen Univ.*, Bd. vii, 1929, pp. 175-205; more particularly pp. 194 ff.

The above proof of the theorem of the square of the diagonal is given in later times in India by Bhāskara II (1150).[1] Bretschneider [2] conjectures that Pythagoras' proof of the theorem was substantially the same. In approving of this hypothesis of Bretschneider, Hankel remarks that "it has no specific Greek colouring but rather reminds of the Indian style." [3] This remark has been accepted by Allman,[4] Gow [5] and Heath.[6] Heath has pointed out another objection against accepting that as Pythagoras' proof, though he admits it to be the "best." [7] This interesting proof was given also by Chang Chun-Ch'ing (*c.* 200 A.D.) in his commentary of the ancient treatise *Chou-pei* (c. 1100 B.C.).[8]

Another plausible hypothesis will be that the ancient Hindus were led to the discovery of the general proof of the theorem of the square of the diagonal in the following way: Let $ABCD$ be a given square. Draw the diagonal AC and cut off AE equal to AC. Construct the square $AEFG$ on AE. Join DE and on it construct the square $DHME$. Complete the construction as indicated in Fig. 69. Now the square $DHME$ is seen to be comprised

1 *Bījagaṇita* (of Bhāskara II), ed. Sudhakara Dvivedi and Muralidhara Jha, Benares, 1927, p. 70.

2 C. A. Bretschneider, *Die Geometrie und die Geometer vor Eukleides*, Leipzig, 1870, p. 82.

3 H. Hankel, *Zur Geschichte der Mathematik in altertum und mittelalter*, Leipzig, 1874, p. 98.

4 G. J. Allman, *Greek Geometry from Thales to Euclid*, Dublin, 1889, p. 37.

5 J. Gow, *A Short History of Greek Mathematics*, Cambridge, 1884, p. 155 f.

6 Heath, *Euclid*, I, p. 355.

7 Heath, *Euclid*, I, p. 355; *Greek Math.*, I, p. 149.

8 Y. Mikami, "The Pythagorean Theorem," *Archiv. Math. Phys.*, XXII (3), 1912, pp. 1-4; *The Development of Mathematics in China and Japan*, Leipzig, 1913, p. 5.

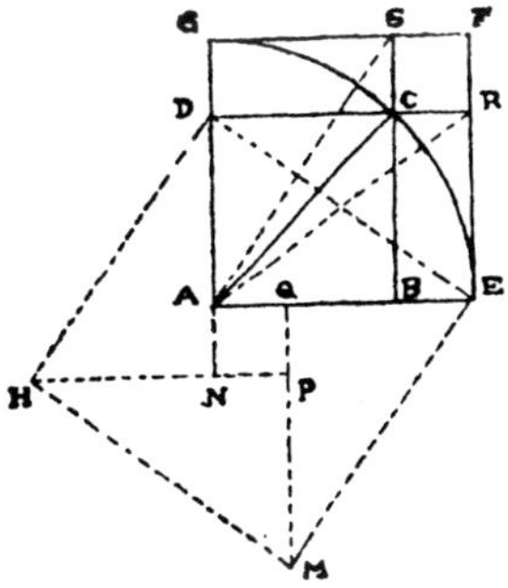

Fig. 69

of four right-angled triangles each equal to DAE and the small square $ANPQ$. This square will be easily recognised to be equal to the square $CRFS$ and triangles equal to the rectangles $AERD$ and $ABSG$. Therefore the square $DHME$ is equal to the sum of the squares $ABCD$ and $AEFG$. Hence the theorem. Constructions like this are necessary in the usual course in the *Śulba*.[1]

Early History

The early history of the theorem of the square of the diagonal, in India, has been very ably treated by Bürk.[2] We have now discovered a few more corroborative evidence of unquestionable value as regards its ancient character. In the *Śulba*, the theorem is found to have been applied very extensively. Even in the construction of a square, a rectangle or a trapezium, is ordinarily presupposed a knowledge of its converse. We may, however, presume for the sake of the best argument on the adverse that in the anterior times, those simple geometrical figures were used to be constructed by methods which would not

[1] For instance see Fig. 57; see also *Paṇḍit*, O.S., X, pp. 46 f.

[2] Bürk, *ZDMG*, LV, pp. 546 ff.

depend in any way on that theorem.[1] The proof of the possibility and existence of such methods, we find also in the *Śulba*. But there are certain other geometrical constructions such as (*i*) the geometrical constructions of $\sqrt{2}$, $\sqrt{3}$, etc., and (*ii*) the transformation of rectangles into squares, for which the theorem of the square of the diagonal is absolutely indispensable. Hence to determine the ancient history of the theorem in India, we shall have to find from the ancient literature of the Hindus the oldest instances of the application of the one or the other of those geometrical constructions. Such instances occur indeed copiously.

Now the doubling of a square, *i.e.*, the geometrical construction of $\sqrt{2}$, is necessary for the construction of the one of the three primarily essential altars of the Vedic sacrifices, *viz.*, the *Dakṣiṇa*. It has been pointed out before that the existence of those three altars is older than the *Ṛgveda* (before 3000 B.C.). Hence the theorem of the square of the diagonal, particularly in its simplest form for the case of the square, is as old as that. In connexion with the construction of the *Rathacakra-citi*, as in the case of the most of the *Kāmya Agni*. one has first to draw a square equal to the primitive and standard *Agni*, the *Caturasra-śyena-cit*, whose area is 7½ square puruṣas and whose form (Fig. 1) consists partly of squares and partly of rectangles. It will be easily seen that the solution of this problem is not possible without the help of the theorem of the square of the diagonal in its general

[1] From the preference which is found to have been given in the *Śulba* to the method of construction of rectangles and trapeziums with the help of the rational rectangle (15, 36, 39) connected with the dimensions of the *Mahāvedi*, which will be noticed in the next chapter, we are convinced that the same method was used to be followed as early as the time of the *Taittirīya* and other *Saṁhitā* (*c*. 3000 B.C.). But here for arguments' sake we shall waive that conviction too.

form.[1] For the *Praüga-citi*, one has again to double that square. Hence the theorem must be as old as the *Taittirīya* and other *Saṁhitā* (*c*. 3000 B.C.) where, it has been pointed out before, occurs the express mention of the construction of the *Kāmya Agni*. In the *Śatapatha Brāhmaṇa*, we are asked to construct a square fourteen times as great as another square of one puruṣa. Again it is required to divide the small square into seven parts and three of the parts will have then to be combined with the larger square to form a new square.[2] From that *Brāhmaṇa*, we come to know of some ancient authorities who used to approve of the construction of a series of *Agni* of the square shape, with areas $1\frac{1}{2}$, $2\frac{1}{2}$,...$6\frac{1}{2}$ square puruṣas.[3] These have been incidentally mentioned by Baudhāyana[4] and Āpastamba.[5] But the *Śatapatha Brāhmaṇa* particularly forbids the construction of such altars. That is, however, quite immaterial for the object we have now in view. It is sufficient and important for us to know that such constructions which clearly depend upon the addition of squares were once in vogue, at any rate in some particular schools, before the time of the *Śatapatha Brāhmaṇa* (*c*. 2000 B.C.). In one of the cases noted above it is necessary to transform a rectangle into a square before it can be added to another. In the *Śrauta-sūtra*, we find copious instances of the geometrical construction of $\sqrt{2}$, $\sqrt{3}$, etc. Some of those instances can be clearly distinguished from all the previous ones

[1] *Cf*. Bürk. *ZDMG*, LV, pp. 549, 553.

[2] *ŚBr*, x. 2.3. 7-14. For further particulars in this connexion see Chap. XII.

[3] *Ibid*, x. 2. 3. 17.

[4] *BŚl*, iii. 318-9.

[5] *ĀpŚl*, viii. 3, 5; xii. 1-2; *ĀpŚr*, xvi. 17.15. In this work, the tradition is expressly attributed to the *Śatapatha Brāhmaṇa*.

inasmuch as in them the application of the theorem of the square of the diagonal has been very clearly mentioned.[1] In the *Baudhāyana Śrauta*, the converse theorem is used for the construction of the *Mahāvedi*.[2] Thus we learn that the theorem of the square of the diagonal really plays a very important part in the science of altar-construction from very early times and in some cases it is, in fact, indispensable.

[1] For example, we take the following from the earliest *Śrauta-sūtra*, namely the *Baudhāyana Śrauta* :

"अथ महावेदिं विमिमीत एतानेव ज्यायसः प्रक्रमान्प्रक्रम्याक्ष्णयामानेन प्रमाय समन्तँ स्पन्दया परितनोति। पृष्ठ्यामातनोत्यथैतमग्निं प्रत्यञ्च यूपावटीयाच्छङ्कोर्विमिमीते पुरुषमात्रेण वेणुना समपक्षपुच्छमरत्निना पक्षौ द्राघीयांसौ भवतः। षड्विधं वा सप्तविधं वा द्वादशविधं वा यावद्विधं वा चेष्यमाणो भवत्यथैनमक्ष्णयामानेन प्रमाय समन्तँ स्पन्दया परितनोति।"—*BŚr*, x. 19.

"वेदिस्तृतीये यजेतेति विज्ञायते। तस्याः सौमिकं मानमेतावदेव नाना। सौमिकात् प्रक्रमात्तृतीयोऽँशः प्रक्रमः स्यात्तेन वेदिं विमिमीत। अपि वा पदाक्ष्णया पार्श्वमानी पदं तिरः पशुमानेन तयोर्यः कर्णसंमितः प्रक्रमः स इष्यते। पदाष्टा नवमस्तदक्ष्णया तयोस्तु यः कर्णसंमितः स प्रक्रमार्थस्तेन मेया सौमिकी वेदिः।"
—*BŚr*, xix. 1.

Compare also *BŚr*, xxvi. 10.

[2] *BŚr*, vi. 22.

CHAPTER X

Rational Rectangles

In the *Śulba-sūtra* we meet with the following rational rectangles :

(a)[1] $3^2+4^2=5^2$,

(i)[2] $9^2+12^2=15^2$,

(ii)[3] $12^2+16^2=20^2$,

(iii)[4] $15^2+20^2=25^2$,

(iv)[5] $72^2+96^2=120^2$.

(b)[6] $5^2+12^2=13^2$,

(i)[7] $15^2+36^2=39^2$,

(ii)[8] $40^2+96^2=104^2$.

(c)[9] $7^2+24^2=25^2$.

(d)[10] $8^2+15^2=17^2$.

(e)[11] $12^2+35^2=37^2$.

1 *BŚl*, i. 49 ; *ĀpŚl*, v. 3.
2 *KŚl P*, verse 31.
3 *ĀpŚl*, v. 3.
4 *ĀpŚl*, v. 3.
5 *MāŚl*, iii. 4-6.
6 *BŚl*, i. 49 ; *ĀpŚl*, v. 4.
7 *BŚl*, i. 49 ; *ĀpŚl*, v. 2, 4 ; *MāŚl*, v. 2-3.
8 *MāŚl*, iii. 3 ; *MaiŚl*.
9 *BŚl*, i. 49.
10 *BŚl*, i. 49 ; *ĀpŚl*, v. 5.
11 *BŚl*, i. 49 ; *ĀpŚl*, v. 5.

Besides these of which the sides and the diagonal are rational *integers*, we find also a few other rectangles whose sides and diagonal are expressible in rational *fractions*:

(*a.v*)[1] $(2\frac{1}{4})^2 + 3^2 = (3\frac{3}{4})^2$

(*a.vi*)[2] $(7\frac{1}{2})^2 + 10^2 = (12\frac{1}{2})^2$

(*b.iii*)[3] $(1\frac{2}{3})^2 + 4^2 = (4\frac{1}{3})^2$

(*b.iv*)[4] $(2\frac{1}{2})^4 + 6^2 = (6\frac{1}{2})^2$

(*b.v*)[5] $(2\frac{1}{12})^2 + 5^2 = (5\frac{5}{12})^2$

(*b.vi*)[6] $(4\frac{1}{6})^2 + 10^2 = (10\frac{5}{6})^2$

(*b.vii*)[7] $(11\frac{1}{4})^2 + 27^2 = (29\frac{1}{4})^2$

(*b.viii*)[8] $(78\frac{1}{3})^2 + (188)^2 = (203\frac{2}{3})^2$

It will be very interesting to know how the early Hindus discovered their rational rectangles. Thibaut observes: "Most likely they discovered that the square on the diagonal of an oblong, the sides of which were equal to three and four, could be divided into twenty-five small squares, sixteen of which composed the square on the longer side of the oblong, and nine of which formed the area of the square on the shorter side. Or, if we suppose a more convenient mode of trying, they might have found that twenty-five pebbles or seeds, which could be arranged

1 *KŚl P*, verse 25f.
2 *MāŚl*, vi.
3 *MaiŚl*.
4 *ĀpŚl*, vi. 6; *MāŚl*, iv; *MaiŚl*.
5 *ĀpŚl*, vi. 7.
6 *ĀpŚl*, vi. 8.
7 *ĀpŚl*, vii. 3.
8 *ĀpŚl*, vi. 5.

in one square, could likewise be arranged in two squares of sixteen and of nine. Going on in that way they would form larger squares, always trying if the pebbles forming one of these squares could not as well be arranged in two smaller squares. So they would form a square of 36, of 49, of 64, etc. Arriving at the square formed by $13 \times 13 = 169$ pebbles, they would find that 169 pebbles could be formed in two squares, one of 144, the other of 25. Further on 625 pebbles could again be arranged in two squares of 576 and 49, and so on." [1]

Thus Thibaut supposes that the Hindus had, starting from a greater square, obtained two smaller ones by division. Bürk on the other hand supposes that they started more likely from a smaller square and found that the new square formed by increasing it was the sum of two smaller ones,—the original square and the square formed by the added portion. This supposition, indeed, tallies more with the procedures found in the *Śulba*. For instance, take the method of construction of the *Śara-rathacakra-cit* described by Baudhāyana. There one has to make, as an auxiliary construction, a square with 225 plus 64 altogether 289 square bricks. Baudhāyana says:

"With these bricks a square is to be formed. The side of a square (first formed) comprises sixteen bricks. Thirty-three bricks will still remain in excess. With them construct the borders (on two sides) completely round." [2]

It will seem strange that instead of being directed to construct the whole square at once, we are told to make at first a square with 256 bricks and then to place the remaining 33 bricks around its two sides. As has been rightly pointed out by the commentator, this rule must be

[1] Thibaut, *Śulvasūtras*, p. 12.

[2] *BŚl*, iii. 191-4.

explained by the fact that the process of construction really began with a square consisting of 4 bricks. Next square was constructed by plFcing 5 squares, along its two sides, so that it contained 9 bricks. Proceeding thus and placing the additional bricks alternately on the north and east sides, and on the south and west sides, a square of 256 bricks was constructed without disturbing the position of symmetry of the altar about the east-west line. But the addition of the remaining 33 bricks displaced this symmetry and for this reason perhaps, Baudhāyana gave that unusual direction.[1]

This explanation is, however, immaterial for the purpose of our immediate object. It is quite sufficient that we have found that a new and larger square was used to be formed from another of smaller size by adding to it a portion in the form of a gnomon. More particularly it is found that a square comprising of 289 square bricks, 17 on each side, was formed from another of 225 bricks, 15 on each side, by the addition of a gnomon consisting of 64 square bricks. Thus is obtained the rational rectangle $15^2 + 8^2 = 17^2$.

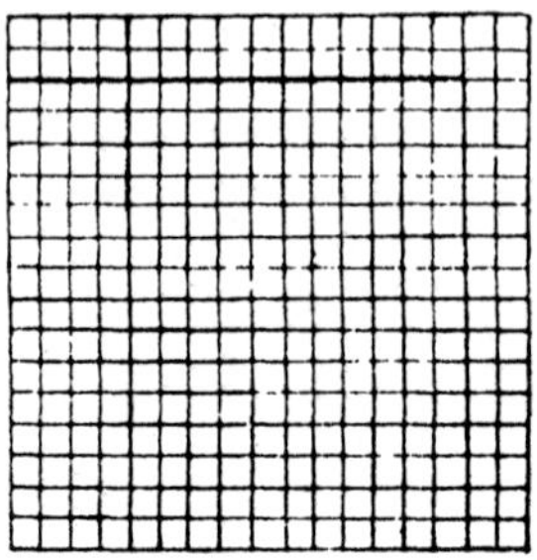

Fig. 70

The whole process appears in a nutshell in the general

[1] Compare Thibaut, *Śulvasūtras*, p. 35f.

rule for the enlargement of a square which has been explained before. It is then simply by noting when the square bricks comprising the gnomon can also be placed

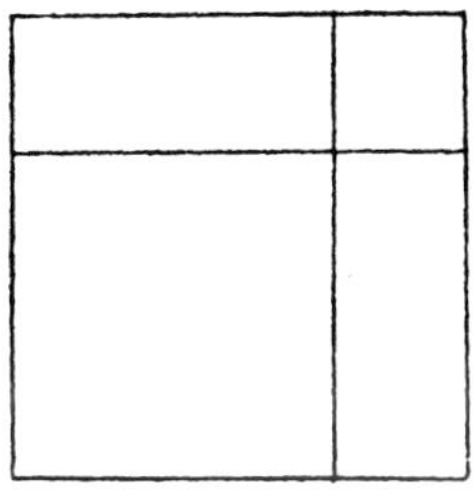

Fig. 71

in the form of a square that the Hindus very likely discovered the rational rectangles. As has been suggested by Treutlein,[1] and followed by Allman [2] and Heath, [3] it was substantially in the same way that Pythagoras discovered the rational right-angled triangles which are now generally associated with his name.

The number of rational rectangles employed in the *Śulba* for the purpose of the construction of altars is found to be few. It will be fewer still—five only—if we consider the independent ones, leaving out the multiples or sub-multiples of them. So it will be naturally asked were the early Hindus aware of other rational rectangles? Had they any general rule for finding any number of rational rectangles?

In dealing with these questions, we shall first refer to two observations of Āpastamba which appear to imply an answer in the negative. After describing four methods

[1] P. Treutlein, *Zeits. f. Math. u. Phys*, xxviii, 1883, Hist.-litt. Abtheilung, pp. 209 ff.

[2] Allman, *Greek Geometry*, pp. 30ff.

[3] Heath, *Euclid*, I, p. 358.

of constructing the *Mahāvedi* in which use has been made of altogether eight rectangles of which the sides and the diagonal are expressible in rational integers, Āpastamba observes: [1]

Etāvanti jñeyāni vedi-viharaṇāni bhavanti.

Thibaut has the reading *vijñeyāni* in the place of *jñeyāni* given by Bürk and also found in my manuscripts. The difference is, however, immaterial. Thibaut translates the passage thus, ' So many " cognizable " measurements of the *vedi* exist.' [2] Bürk renders it as ' There are so many " recognizable " (erkennbare) constructions of the *vedi*,' [3] and has thus closely followed Thibaut. The latter further observes: " That means: these are the measurements of the *vedi* effected by oblongs, of which the sides and the diagonal can be known, *i.e.*, can be expressed in integral numbers." Bürk is of the same opinion.

The other observation of Āpastamba which we shall refer to is

tābhirjñeyābhiruktaṃ viharaṇaṃ

and it occurs at the end of his enunciation of the theorem of the square of the diagonal of a rectangle.[4] Bürk renders it thus: " The construction (in i. 2 and 3) has been taught by means of (the application of) these (the *akṣṇayārajju*, *pārśvamānī* and *tiryaṅmānī* of a rectangle) —of course by means of such as are " recognizable " (*i.e.*, which can be expressed in integral numbers)."

Thus according to the supposition of Thibaut which is accepted also by Bürk, the qualifying words *jñeyāni* and *jñeyābhiḥ* imply only those rectangles of which the sides and the diagonal can be expressed in rational integers.

1 *ĀpŚl*, v. 6.

2 Thibaut, *Śulvasūtras*, p. 12.

3 *ZDMG*, LVI, p. 341; compare also p. 329; LV, p. 560, fn. 3.

4 *ĀpŚl*, i. 4.

This supposition is doubtless wrong. For in the first expression the word *jñeyāni* very clearly qualifies *vedi-viharaṇāni* or " the methods of construction of the *vedi*," the word *vedi* undoubtedly referring to the *Mahāvedi* mentioned in the foregoing rules. This is further confirmed by the word *etāvanti* meaning " these," " so many," the reference being to the four methods of constructing the *Mahāvedi* described by Āpastamba. Where is then the ground to suppose, as has been done by Thibaut and Bürk, that *jñeyāni* implies the rectangles employed for the purpose of those methods ? In the second passage the word *viharaṇaṃ* (" construction " or " the method of constructions ") clearly [1] refers to the methods of construction described in the two foregoing rules, *viz.*, *ĀpŚl*, i. 2 and 3, and the word *jñeyābhiḥ* to the rational rectangles used therein. This has been admitted by Bürk also. Now the methods referred to are those for the construction of squares and rectangles of given sides, and they primarily depend upon drawing right-angled triangles (or rectangles) having a given side, or more particularly, having a side equal to a side of the required figure. The sides and the diagonal of the rectangles are stated to be $(a, 5a/12, 13a/12)$ and $(a, 3a/4, 5a/4)$. It is only when a is a multiple of 12 (in the first case) or a multiple of 4 (in the second case) that the sides and the diagonal of the rectangles employed for the purpose of the construction of altars, will be expressible in rational integers, other-

[1] Compare what Āpastamba says : *uktam* "said," "stated" or " described ; " meaning that the method of construction has been described before. This can be further confirmed by referring to what Āpastamba has written in the rule just preceding the one under discussion (i. 3). After stating the dimensions of the sides and the diagonal of the rectangle to be used, *viz.*, $(a, 3a/4, 5a/4)$, Āpastamba observes *vyākhyātaṃ viharaṇam*, " the method of construction has been (already) described (in the preceding rule)."

wise not. Indeed in actual practice Āpastamba has employed four particular cases of the first, *viz.*, $(2\frac{1}{2}, 6, 6\frac{1}{2})$, $(2\frac{1}{12}, 5, 5\frac{5}{12})$, $(4\frac{1}{6}, 10, 10\frac{5}{6})$ and $(11\frac{1}{4}, 27, 29\frac{1}{4})$, which have fractional sides and diagonals.[1]

Again for the purpose of the construction of altars, Āpastamba has used even such rectangles in which the sides and the diagonal cannot be expressed in terms of rational numbers.[2] So the interpretation of the words *jñeyāni* and *jñeyābhiḥ*, as supposed by Thibaut and Bürk, is absolutely untenable.

Having detected that this interpretation is open to such a serious objection, Heath modifies it and says: "But the words (*etāvanti jñeyāni*, etc.) also imply that the theorem of the square of the diagonal is also true of other rectangles not of the "recognisable" kind, that is rectangles in which the sides and the diagonal are not in the ratio of integers; this is indeed implied by the constructions for $\sqrt{2}$, $\sqrt{3}$, etc., up to $\sqrt{6}$ (*cf.* ii. 2, viii. 5)."[3] But he would still presume that the remark implies that Āpastamba knew of no other rational rectangles that could be employed.

Let us next turn to find how the passages in question have been explained by the orthodox commentators. Amongst them, the most elaborate explanation of the expression *tabhirjñeyābhir*, etc., is found to be that of Karavindasvāmī. According to him, the word *jñeya* implies that variety of quadrilaterals in which of the sides and the diagonal any two being given, the third "can be known" with the help of the theorem of the square of

[1] *ĀpŚl.* vi. 6-8, vii. 3.

[2] *Vide* supra, p. 121. Further in two other methods of construction Āpastamba has used the isosceles right-angled triangle (or a square) a, a, $a\sqrt{2}$ (ii. 1, ix. 3).

[3] Heath, *Euclid*, I, p. 368.

the diagonal; or it denotes the rectangles which "can be conceived in mind," the sides and the diagonal of them being expressible in terms of commensurable quantities.[1] The first explanation is given also by Sundararāja and the alternative one by Kapardisvāmī. Sundararāja is silent about the true import of the word *jñeyāni* in *etāvanti jñeyāni*, etc. In the opinion of Karavindasvāmī and Kapardisvāmī, it implies those rectangles in which the sides and the diagonal "can be known in terms of numbers which are rational (*śuddhamūla*, lit. 'which are perfect roots')."[2]

These interpretations appear to me to be as unnatural and forced as those of Thibaut and Bürk. Every one of

[1] Karavindasvāmī writes:

"प्रमाणकारपार्श्वमानीतिर्यङ्मान्यक्ष्णयारज्जूनामन्यतरयोः परिमाणज्ञाने अन्यतरा ज्ञातुं शक्यते ता ज्ञेयाः। तत् कथं ··· अक्ष्णयारज्ज्वः फलभूतक्षेत्रात् पार्श्वमानीफलभूतक्षेत्रे च शोधिते शिष्टक्षेत्रस्य करणी तिर्यङ्मानीति ज्ञातुं शक्यते। एवमेताः ज्ञेयाः। या एवं ज्ञेयाः ताभिरेताभिरुक्तं विहरणमित्यर्थः। एवमुत्तरस्य विहरणस्य समाधिहेतुत्वमवगन्तव्यमित्यर्थः। अथवा ज्ञेयाभिः परिकल्पयितुमुचिताभिः पूर्वोक्तविहरणं कर्त्तव्यमित्ययमर्थः।"

Sundararāja says:

"आसां द्वयोर्ज्ञातयोस्तृतीया ज्ञातुं शक्यते। यथा, पार्श्वमानीतिर्यङ्मान्योर्ज्ञातयोस्ते पृथग् वर्गयित्वा संयोज्य तद्वर्गमूलमक्ष्णयारज्जुः तथा पार्श्वमान्यक्ष्णयारज्ज्वोर्ज्ञातयोरक्ष्णयारज्जुवर्गात् पार्श्वमानीवर्गं विशोध्य शिष्टस्य मूलं तिर्यङ्मान्येवं तिर्यङ्मानीवर्गं विशोध्य पार्श्वमानी। एवं ताभिर्ज्ञेयाभिः पूर्वोक्तं विहरणं ···।"

Kapardisvāmī has:

"ज्ञेयाभिर्ज्ञातुं शक्याभिर्मनसा परिकल्पिताभिः।"

[2] Karavindasvāmī says:

"ज्ञेयानि शुद्धमूलतया ज्ञातुं शक्यानि।"

Kapardisvāmī says:

"एतावन्त्येव शुद्धमूलानि ज्ञातुं शक्यानि वेदिविहरणानि भवन्ति। अन्ये तु अशुद्धमूलाः कल्पयितुमशक्याः। तस्मादेतावन्तीत्येवावधार्यन्ते।"

these interpretations is putting a construction on the text which it does not seem to bear. At any rate those interpretations have been made irrespective of the sense in which the words very closely related to *jñeya* have been used in the *Śulba* and of the nature and spirit of these works.

The question of the rationality or irrationality of the sides and the diagonal of the rectangles used in the construction of altars can arise only when we begin to think of them in terms of numerical quantities. But the *Śulba* deals truly with geometrical construction and not with numerical calculation. In commenting upon *jñeya* from the standpoint of numerical representation, the commentators, ancient as well as modern, have fallen into an error.

I think that the word *jñeya* should be explained quite differently. Its literal significance is, as has been explicitly stated by the commentators, "can be known" (*jñātuṃ śakyate*). It comes from the same root *jñā*, as the word *vijñāyate*, meaning "is known." Hence both the words should be explained so as to exhibit the same relation. Now on many occasions in the *Śulba* in connexion with the statements of the measurements of altars, the description of the methods of constructing them, etc., we find the remark...*iti*...*vijñāyate*.[2] It implies that such and such thing "is known from the ancient holy scriptures." Indeed those things can be actually traced therein. I think the word *jñeya* also should be explained as implying a reference to the same ancient scriptures. Thus the expression *tābhirjñeyābhir*,

[1] It is noteworthy that this is one of the arguments applied by Bürk against the interpretation of Sundararāja.

[2] For instance see *BŚl*, i. 65, 71, 76, 79, etc.; *ĀpŚl*, iv. 1, 3, 5; v. 1, 8, 10; viii. 1, 4, 7, etc.

etc., means "the method of construction (of the altars) has been described by means of those rectangles that can be known (from the ancient scriptures)." The other expression *etāvanti jñeyāni*, etc., will then mean "these only are the methods of construction of the (*Mahā-*)*vedi* which can be known (from the ancient scriptures)."

We shall now take up the question whether the ancient Hindus had any general method of finding rational triangles. Of course there is not found in the *Śulba* any rule devoted particularly to the definition of such a method. So whatever we shall say on the point will consequently be by way of inference, more or less conjectural. However there are good reasons to believe that the Hindus knew of general formulas for finding rational rectangles.

We have already pointed out how the ancient Hindus presumably discovered the rational rectangles which are found in the *Śulba* by noting when the square bricks comprising the gnomon added to a square might themselves be arranged again in the form of a square. Observing that the gnomon of one square brick depth put round a square formed with n^2 such bricks, consists of $2n+1$ bricks, they would have only to make a square with $2n+1$ bricks.[1]

If we suppose that $2n+1 = m^2$,

we obtain $n = \frac{1}{2}(m^2-1)$;

and therefore $n+1 = \frac{1}{2}(m^2+1)$.

It follows that

$$m^2 + \left(\frac{m^2-1}{2}\right)^2 = \left(\frac{m^2+1}{2}\right)^2. \qquad \ldots \quad \text{(A)}$$

This formula follows, indeed, more directly from a special rule of Kātyāyana for finding the sum of a number of equal squares. If n be the number of equal squares of

[1] Compare Müller, *loc. cit.*, pp. 202 f.

sides equal to a each to be combined into one, then, the rule says,

$$na^2 = \left(\frac{n+1}{2}\right)^2 a^2 - \left(\frac{n-1}{2}\right)^2 a^2.$$

Putting m^2 for n, we get

$$m^2a^2 + \left(\frac{m^2-1}{2}\right)^2 a^2 = \left(\frac{m^2+1}{2}\right)^2 a^2. \quad \ldots \quad (\acute{A})$$

In particular, taking $a=1$, we get at once the formula (A). If the sides and the diagonal of the rectangles are to be integral as well as rational, m must be odd.

With the help of this formula would be obtained the following rational rectangles mentioned in the *Śulba*: (3, 4, 5), (5, 12, 13) and (7, 24, 25). Indeed it will give all those rational rectangles in which the difference between the greater side and the diagonal is 1.

There are also found other rational rectangles, *viz.*, (8, 15, 17) and (12, 35, 37), which could not be obtained from the formula (A). The characteristic of them is that the difference between the greater side and the diagonal is 2. They could be obtained from the formula

$$(2m)^2 + (m^2-1)^2 = (m^2+1)^2 \quad \ldots \quad (B)$$

which is derivable from (A) by doubling the side of each square or from ($\acute{A}$) by putting $a=2$. But they were more likely obtained first by observing when the gnomon of two square bricks of breadth put round a square, could be rearranged in the form of a square. If the original square contain n^2 bricks, the gnomon will consist of $4n+4$ bricks.

If we suppose that $4n+4 = m^2$,

we obtain $n = \frac{1}{4}(m^2-4)$;

and therefore $n+2 = \frac{1}{4}(m^2+4)$.

It follows that

$$m^2 + \left(\frac{m^2-4}{4}\right)^2 = \left(\frac{m^2+4}{4}\right)^2.$$

Substituting $2m$ for m in this, we easily obtain the formula (B).

Proceeding in the same way they could deduce from the general rule for the enlargement of a square, a still general formula for finding rational rectangles. It has been stated in that rule that the gnomon of n bricks depth put round a square of p^2 bricks will contain $2pn+p^2$ bricks. Supposing

$$2pn+n^2 = m^2,$$

we obtain
$$p = \frac{1}{2n}(m^2-n^2);$$

$$p+n = \frac{1}{2n}(m^2+n^2).$$

It follows that

$$m^2 + \left(\frac{m^2-n^2}{2n}\right)^2 = \left(\frac{m^2+n^2}{2n}\right)^2,$$

or
$$(2mn)^2 + (m^2-n^2) = (m^2+n^2). \quad \ldots \quad \text{(C)}$$

where m, n are any two rational integers.

This formula follows also from the method of the transformation of a rectangle into a square, which is commonly found in all the works on the *Śulba*. If p, q be the length and breadth of the rectangle to be transformed, it has been stated that the equivalent square will be given by the difference of the two squares

$\left(\frac{p+q}{2}\right)^2$ and $\left(\frac{p-q}{2}\right)^2$. Thus

$$pq + \left(\frac{p-q}{2}\right)^2 = \left(\frac{p+q}{2}\right)^2.$$

Substituting m^2, n^2 for p, q respectively, we get

$$m^2n^2 + \left(\frac{m^2-n^2}{2}\right)^2 = \left(\frac{m^2+n^2}{2}\right)^2. \qquad \ldots \quad (C')$$

Proclus (450 A.D.) has attributed the formula (A) to Pythagoras (*c.* 540 B.C.), and the formula (B) to Plato (*c.* 375 B.C.). The formula (C) or (C′) follows from Euclid's *Elements*, II. 6.

The early Hindus recognised that fresh rational rectangles can be derived from a known one by multiplying or dividing its sides and diagonal by any rational quantity. In other words, they found that if (p, q, r) be a rectangle, so that

$$p^2 + q^2 = r^2,$$

then another will be (lp, lq, lr), where l is any rational number, integral or fractional. Āpastamba has, indeed, derived certain new rational rectangles in the same way. He has, however, put the result thus:[1]

If $$\alpha^2 + \beta^2 = \gamma^2,$$

then $$(\alpha + n\alpha)^2 + (\beta + n\beta)^2 = (\gamma + n\gamma)^2,$$

where n is an arbitrary rational number.

The rational rectangle (15, 36, 39) perhaps deserves more than a passing notice. It could of course be derived from the rational rectangle (5, 12, 13) by multiplying its sides and diagonal by 3. This relation has, indeed, been expressly admitted by Āpastamba.[2] But it was probably obtained first independently, thinks Bürk,[3] as an instance of a rational rectangle in which the difference between the greater side and the diagonal would be 3. This will probably account for Baudhāyana's enumerating it separately along with (5, 12, 13). The early Hindus,

[1] *ĀpŚl*, v. 3-4.
[2] *ĀpŚl*, v. 4.
[3] *ZDMG*, LV, p. 571.

particularly those belonging to the Āpastamba school, appear, however, to have special regard for the rational rectangle (15, 36, 39). So the method of constructing several *vedis* has been described in the *Śulba* of this school with particular reference to this rectangle. For instance, take the case of the *Nirūḍhapaśubandha-vedi*. It is of the shape of an isosceles trapezium whose measurements are known from the ancient scriptures to be: face = the yoke of a cart (=86 aṅgulis); altitude = the pole (=188 aṅgulis) and base = the axle of the cart (= 104 aṅgulis).[1] As regards the process of its construction, Āpastamba says:

"This has been described (in connexion with the construction of the *Saumikī-vedi*) by means of one cord. Having taken it by the mark at 15, fix the two western corners by means of half the axle and the eastern corners by means of half the yoke."[2]

Here it is clear that the altitude (=188 aṅgulis) of the *vedi* is supposed to be divided into 36 parts, so that one part will be equal to $5\frac{2}{9}$ aṅgulis. On taking this for a new unit of pada or prakrama, in terms of it the altitude of the *Nirūḍhapaśubandha-vedi* will contain 36 padas or prakramas. So the rational rectangle (15, 36, 39) can be applied to construct it, as in the Method I, page 64 (Fig. 28). The spatial magnitudes of the rational rectangle employed will truly be ($78\frac{1}{3}$, 188, $203\frac{2}{3}$) in terms of the usual unit aṅguli. But by supposing the unit to be one of $5\frac{2}{9}$ aṅgulis, Āpastamba represents it as (15, 36, 39). Similarly in constructing several other altars, by a suitable change of the length of the unit of linear measure, Āpastamba always represents the sides and the diagonal of the rational rectangle employed by (15, 36,

[1] *ĀpSl*, vi. 3, 5; *ĀpŚr*, vii. 3. 7 f.

[2] *ĀpSl*, vi. 4; compare Bürk's notes on it.

39).[1] Actually they are particular cases of the rectangle $(a, 5a/12, 13a/12)$, a being the length of the altitude of the *vedi* under construction.

Now it will be very naturally asked, why this preferential liking for the particular rational rectangle (15, 36, 39), on the part of Āpastamba? The true answer will be not simply because, as Bürk seems to think, that the construction of the most important *vedi*, namely the *Mahāvedi* or the *Saumikī-vedi*, depended on it. For Āpastamba has described as many as four methods for the construction of the same *vedi*. But also because it was employed in the most ancient method of construncting the *Mahāvedi* and so had acquired a special sanctity by a long scriptural tradition. We have also seen the similar orthodox predilection of Āpastamba for the primitive methods of high scriptural antiquity in the matter of the construction of the *Caturasra-śyenacit* by means of the bamboo-rod.

Now tracing the early history of the rational rectangle (15, 36, 39), we find it first in the *Taittirīya Saṁhitā* (c. 3000 B.C.)[2] in connexion with the *Mahāvedī*. It has then reappeared in the *Kāṭhaka Saṁhitā*,[3] *Maitrāyaṇī Saṁhitā*,[4] *Kapiṣṭhala Saṁhitā*,[5] and *Śatapatha Brāhmaṇa*.[6] It should perhaps be noted that in these works only the sides (15, 36) have been expressly mentioned, but not the diagonal 39. This non-mention of the diagonal has led some modern writers to suspect if the property $15^2+36^2=39^2$ was at all known in the time

1 *ĀpŚl*, vi. 6-8.

2 *TS*, vi. 2. 4. 5.

3 *KṭS*, xxv. 4.

4 *MaiS*, iii. 8. 4.

5 *KapS*, xxxviii. 6.

6 *ŚBr*, iii. 5. 1. 1ff.; x. 2. 3. 4.

of the *Saṁhitā* and *Brāhmaṇa*.[1] But such a suspicion seems to be quite unwarranted. For even in later times Baudhāyana has not mentioned the diagonal of the rational rectangles enumerated by him. From a thorough discussion of the point, Bürk concludes: "After all these no doubt can exist regarding the fact that the rational right-angled triangle with perpendicular sides 15 and 36 was really known in the time of the *Taittirīya Saṁhitā* and the *Śatapatha Brāhmaṇa* and was employed in the construction of the *Saumikī-vedi* as in the *Āpastamba Śulba Sūtra* v. 1 and 2." [2] Such was also the opinion of Cantor.[3] Their conclusion will be further corroborated by what has been shown just above about the special sanctity acquired by this rational rectangle amongst the followers of the Āpastamba school. Our interpretation of the word *jñeyābhiḥ* occurring at the end of Āpastamba's enunciation of the theorem of the square of the diagonal of a rectangle (pp. 132 f.) also will lead one strongly to the same conclusion.

[1] For instance, see Keith, *Journ. Roy. Asiat. Soc.*, 1909, pp. 590 f.; 1910, pp. 519 f.

[2] *ZDMG*, LV; pp. 555 f.

[3] Cantor, *Geschichte*, I, pp. 598 ff.

CHAPTER XI

SQUARING THE CIRCLE

To transform a square into a circle.

Baudhāyana says:

" If you wish to circle a square, draw half its diagonal about the centre towards the east-west line ; then describe a circle together with the one-third of that which lies outside (the square)." [1]

The same method has been taught also by Āpastamba [2] and Kātyāyana.[3]

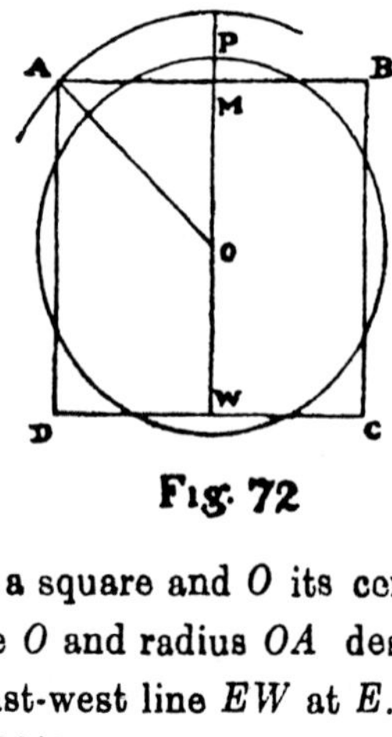

Fig. 72

Let $ABCD$ be a square and O its central point. Join OA. With centre O and radius OA describe a circle intersecting the east-west line EW at E. Divide EM at P such that $PM = EM/3$. Then with centre O and radius

[1] " चतुरश्रं मण्डलं चिकीर्षन्नक्ष्णयार्धं मध्यात् प्राचीमभ्यापातयेद्यदतिशिष्यते तस्य सह तृतीयेन मण्डलं परिलिखेत् । "—*BSl*, i. 58.

[2] " चतुरश्रं मण्डलं चिकीर्षन्मध्यात् कोट्यां निपातयेत् पार्श्वतः परिकृष्यातिशयतृतीयेन सह मण्डलं परिलिखेत् । सानित्या मण्डलम् । यावद्धीयते तावदागन्तु । "—*ĀpSl*, iii, 2.

[3] *KŚl*, iii. 13.

OP describe a circle. This circle will be nearly equal in area to the given square $ABCD$.

Let $2a$ denote a side of the given square and r the radius of the circle equivalent to it ; that is, $AB=2a$. $OP=r$. Then

$$OA = a\sqrt{2},$$

$$\text{and } ME = (\sqrt{2} - 1)a.$$

$$\text{Hence} \qquad r = a + \frac{a}{3}(\sqrt{2} - 1),$$

$$= \frac{a}{3}(2 + \sqrt{2}).$$

Now according to the *Śulba*,[1]

$$\sqrt{2} = 1 + \frac{1}{3} + \frac{1}{3.4} - \frac{1}{3.4.34}$$

$$= \frac{577}{408},$$

$$= 1{\cdot}4142156...$$

$$\text{Therefore} \qquad r = a \times 1{\cdot}1380718...$$

The area of the transformed circle, employing the value $\pi=3{\cdot}14159$, will be $4{\cdot}068987 \times a^2$ whereas the area of the given square is $4a^2$. Hence the former result is too large. On the degree of equivalence of the area of the given square and of the circle into which it is transformed by the above rule, there is a noteworthy observation of Āpastamba. He remarks,

sānityā maṇḍalaṃ :yāvaddhīyate tāvadāgantu.[2]

According to the commentator Kapardisvāmī, the first word of this passage is a conjoint compound of the two

[1] *Vide infra*, p. 189.

[2] *ĀpSl*, iii. 2.

words *sā* and *anityā*. So that the above should be rendered as, "It is an inexact (*anityā*) (method of construction); by as much the circle falls short, so much comes in." He has been followed in this respect by Sundararāja. But the commentator Karavindasvāmī thinks that the correct reading of the passage will be *sā nityā maṇḍalaṃ,'* etc. According to this reading, the observation appears to imply that the method of construction is an exact (*nityā*) one. But this commentator further thinks that the method has been called "exact" in a relative sense inasmuch as it yields a result more accurate than that obtained by any other method of circling the square known in the Āpastamba school. So he too finally comes to the same point of view as that Kapardisvāmī and Sundararāja.

Thibaut accepts as correct the reading of the text as given by Karavindasvāmī but has discarded his further explanation of the matter. So he renders the passage as: "this line gives a circle exactly as large as the square; for as much as there is cut off from the square (*viz.*, the corners of the square), quite as much is added to it (*viz.*, the segments of the circle, lying outside the square)."[1] Bürk has closely followed Thibaut in this respect. The interpretation of Kapardisvāmī has been criticised by Thibaut thus: "But I am afraid we should not be justified in giving to Āpastamba the benefit of this explanation. The words 'yāvaddhīyate, etc.' seem to indicate that he was perfectly satisfied with the accuracy of his method and not superior, in this point, to so many circle-squarers of later times. The commentator who, with the mathematical knowledge of his time, knew that the rule was an imperfect one, preferred very naturally the interpretation

[1] Thibaut, *Śulbasūtras*, p. 26.

which was more creditable to his author."[1] Kapardisvāmī explains that the remark *yāvaddhīyate*, etc., implies only a nearer approximation but not exact equality.

Now it may be pointed out that a similar remark, *viz.*,

eṣānityā caturasra-karaṇī

has been made by Āpastamba as regards his method for squaring the circle and which has been delivered by him immediately following the above one.[2] The same remark is found also in the corresponding rule of Baudhāyana.[3] Here Thibaut entirely agrees with the commentators in breaking up the first word *eṣānityā* into *eṣā anityā*. But Bürk falls out from him and takes the reading to be *eṣā nityā caturasra-karaṇī*. Thus he has kept up the consistency of his opinion.

It is truly very difficult to conjecture now correctly what was really implied by Āpastamba by that remark, whether he held that method of circling a square was an exact or an inexact one. We may be, however, sure to this exent that the interpretation of the orthodox commentators cannot be brushed aside as unlikely as is supposed by Thibaut.

To transform a circle into a square.

Baudhāyana says:

"If you wish to square a circle, divide its diameter into eight parts; then divide one part into twenty-nine parts and leave out twenty-eight of these; and also the sixth part (of the preceding sub-division) less the eighth part (of the last)."[4]

[1] *Ibid*, p. 27.

[2] *ĀpŚl*, iii. 3.

[3] *BŚl*, i. 60.

[4] "मण्डलं चतुरश्रं चिकीर्षन्विष्कम्भमष्टौ भागान् कृत्वा भागमेकोनत्रिंशधा विभज्याष्टाविंशतिभागानुद्धरेद्भागस्य च षष्ठमष्टमभागोनम्।"—*BŚl*, i. 59.

Thus if $2a$ denote the side of a square equivalent to a circle of diameter d, then

$$2a = \frac{7d}{8} + \left[\frac{d}{8} - \left\{\frac{28d}{8.29} + \left(\frac{d}{8.29.6} - \frac{d}{8.29.6.8}\right)\right\}\right]$$

$$\text{or} \quad 2a = d - \frac{d}{8} + \frac{d}{8.29} - \frac{d}{8.29}\left(\frac{1}{6} - \frac{1}{6.8}\right).$$

Since $d = 2r$, where r is the radius of the circle,

$$a = r - \frac{r}{8} + \frac{r}{8.29} - \frac{r}{8.29.6} + \frac{r}{8.29.6.8}.$$

This result was probably obtained from the previous one by inversion

$$r = \frac{a}{3}(2 + \sqrt{2}).$$

$$\text{Therefore} \qquad 2a = \frac{3}{2 + \sqrt{2}}\, d.$$

Substituting the value of $\sqrt{2}$, *viz.*, 577/408, we have

$$2a = \frac{1224}{1393} d.$$

Thibaut supposes that Baudhāyana then proceeded in the following way: "One-eighth of 1393 = $174\frac{1}{8}$; this multiplied by 7 = $1218\frac{7}{8}$. Difference between $1218\frac{7}{8}$ and 1224 = $5\frac{1}{8}$. Dividing 174 (Baudhāyana takes 174 instead of $174\frac{1}{8}$, neglecting the fraction as either insignificant or, more likely, as inconvenient) by 29 we get 6 ; subtracting from 6 its sixth part we get 5 and adding to this the eighth part of the sixth part of six, we get $5\frac{1}{8}$. In other words:

$$1224 = \frac{7}{8} + \frac{1}{8.29} - \frac{1}{8.29.6} + \frac{1}{8.29.6.8} \text{ of } 1393$$

(due allowance made for the neglected $\frac{1}{8}$)."[1]

[1] Thibaut, *Śulbasūtras*, p. 28.

In the opinion of Cantor the series was probably obtained thus:[1]

$$\frac{1224}{1393}=\frac{7}{8}+\frac{1}{8.29}-\frac{1}{8.29.6}+\frac{1}{8.29.6.8}-\frac{41}{8.29.6.8.1393}$$

The last term is nearly $\frac{1}{34}$ of the term preceding it and so may be neglected as being comparatively small. Hence, we get

$$2a=\frac{7d}{8}+\frac{d}{8.29}-\frac{d}{8.29.6}+\frac{d}{8.29.6.8}.$$

Müller conjectures the procedure adopted to have been as follows:[2]

$$2a=\frac{3}{2+\surd 2}\,d=\frac{3\surd 2}{2\surd 2+2}\,d=\frac{3}{2}\cdot\frac{\surd 2}{1+\surd 2}\,d,$$

$$=\frac{3}{2}\cdot\frac{17-1/34}{29-1/34}=\frac{51-3/34}{58-2/34},$$

since $\surd 2=\frac{17}{12}-\frac{1}{12.34}$. Now

$$\frac{51-3/34}{58-2/34}=1-\frac{7+1/34}{58-2/34}.$$

$$\frac{7+1/34}{58-2/34}=\frac{1}{8}\cdot\frac{56+8/34}{58-2/34}=\frac{1}{8}\left(1-\frac{2-10/34}{58-2/34}\right),$$

$$\frac{2-10/34}{58-2/34}=\frac{1}{29}\cdot\frac{2-10/34}{2-2/34.29}=\frac{1}{29}\left(1-\frac{10/34-2/34.29}{2-2/34.29}\right)$$

$$\frac{10/34-2/34.29}{2-2/34.29}=\frac{5-1/29}{34-1/29}=\frac{1}{6}\cdot\frac{30-6/29}{34-1/29},$$

$$=\frac{1}{6}\left(1-\frac{4+5/29}{34-1/29}\right),$$

[1] Cantor, *Geschichte*, I, p. 643.

[2] C. Müller, *loc. cit.*, pp. 187 f.

$$\frac{4+5/29}{34-1/29} = \frac{1}{8} \cdot \frac{32+40/29}{34-1/29} = \frac{1}{8}\left(1 - \frac{2-41/29}{34-1/29}\right),$$

Thus we have

$$\frac{3}{2+\sqrt{2}} = 1 - \frac{1}{8} + \frac{1}{8.29} - \frac{1}{8.29}\left(\frac{1}{6} - \frac{1}{6.8}\right)$$

$$- \frac{1}{8.29.6.8} \cdot \frac{2-41/29}{34-1/29}.$$

The last term may be neglected as being too small. Hence

$$2a = d - \frac{d}{8} + \frac{d}{8.29} - \frac{d}{8.29}\left(\frac{1}{6} - \frac{1}{6.8}\right).$$

Alternative Method.

Another method of squaring a circle has been taught by Baudhāyana, Āpastamba and Kātyāyana. It has been explicitly admitted by all of them that this method yields only a gross (*anityā*) value.

" Or else divide (the diameter) into fifteen parts and remove two (of them). This is the gross (value of a) side of the (equivalent) square."[1]

That is to say,

$$2a = d - \frac{2}{15} d,$$

or, $$a = r - \frac{2}{15} r.$$

The *rationale* of this formula seems to be this:[2] Draw the square *ABCD* circumscribing the circle and also the square *A′B′C′D′* inscribed within it.

[1] *BŚl*, i. 69. See also *ĀpŚl*, iii. 3; *KŚl*, iii. 14.

[2] Compare Müller, *loc. cit.*, p. 182,

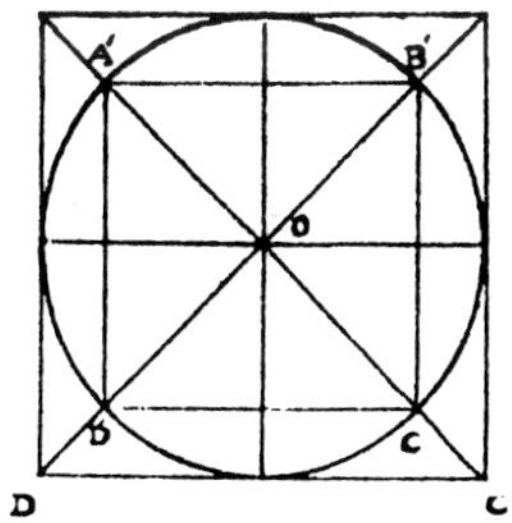

Fig 73

Then apparently the area of the circle will be smaller than the area of the square $ABCD$ $(= 4r^2)$ and greater than the area of the square $A'B'C'D'$ $(= 2r^2)$; that is

$$4r^2 > \text{Area of the circle} > 2r^2.$$

An obvious approximation will be

$$\text{Area of the circle} = \frac{4r^2 + 2r^2}{2} = 3r^2.$$

If $2a$ denote a side of the square equivalent in area to the circle, we shall have approximately

$$4a^2 = 3r^2,$$

$$\text{or,} \quad a = \frac{\sqrt{3}}{2} r.$$

But, it will be shown later on[1]

$$\sqrt{3} = 1 + \frac{2}{3} + \frac{1}{15} = \frac{26}{15},$$

up to the second order of approximation. Therefore

$$a = \frac{13}{15}r = r - \frac{2}{15} r.$$

[1] *Infra*, p. 195.

Dvārakanātha's Corrections.

Dvārakanātha Yajvā has criticised with the help of specific examples the *Śulba* method of squaring a circle and its converse, as yielding only approximate results. He has then proposed the following corrections to the ancient formulae :[1]

$$(i) \quad r = \left\{a + \frac{a}{3}(\sqrt{2} - 1)\right\}\left(1 - \frac{1}{118}\right).$$

$$(ii) \quad 2a = \left\{d - \frac{d}{8} + \frac{d}{8.29} + \frac{d}{8.29}\left(\frac{1}{6} - \frac{1}{6.8}\right)\right\} \times \left(1 + \frac{1}{2}\cdot\frac{3}{133}\right).$$

By the formula (*i*) the area of the transformed circle will be nearly equal to $4{\cdot}000344 \times a^2$, the value of π being taken to be 3·14159.

Value of π.

The above rules of the *Śulba* for squaring a circle and *vice versâ*, will work out the following values of π :

$$(1) \quad \pi = \frac{4}{\left\{1 + \frac{1}{3}(\sqrt{2} - 1)\right\}^2} = 3{\cdot}0883...$$

$$(2) \quad \pi = 4\left(1 - \frac{1}{8} + \frac{1}{8.29} - \frac{1}{8.29.6} + \frac{1}{8.29.6.8}\right)$$

$$= 3{\cdot}0885$$

$$(3) \quad \pi = 4(1 - \tfrac{2}{15})^2 = 3{\cdot}004.$$

It will be noticed that none of these values are fairly accurate, as according to modern calculation $\pi = 3{\cdot}14159...$

[1] *Paṇḍit*, O.S., Vol X, p. 21.

In one instance Baudhāyana[1] has employed the value of $\pi=3$. A better value of π is found in the *Mānava Śulba*, which states that a square of two by two cubits is equivalent to a circle of radius 1 cubit and 3 aṅgulis.[2] Whence

$$(4) \quad \pi = 4(\tfrac{8}{9})^2 = 3{\cdot}16049$$

This value was given before by the Egyptian Ahmes (*c.* 1500 B.C.). With Dvārakanātha's corrections we have

$$\pi = \frac{4}{+\frac{1}{3}(\sqrt{2}-1)\}^2}\left(\frac{118}{117}\right)^2 = 3{\cdot}141109\ldots$$

$$\pi = 4\left(1-\frac{1}{8}+\frac{1}{8{\cdot}29}-\frac{1}{8.29.6}+\frac{1}{8.29.6.8}\right)^2 \times\left(1+\frac{1}{2}\cdot\frac{3}{133}\right)^2$$

$$= 3{\cdot}157991\ldots$$

Early History.

It will be interesting to know the early history of the problem of the squaring of the circle and of its converse, the circling of the square, in India. Their origin dates, as has been noted before, earlier than the time of the *Ṛg-veda* (before 3000 B.C.). That was in connection with the construction of the three primarily essential sacrificial altars of the Vedic Hindus, namely the *Gārhapatya*, *Āhavanīya* and *Dakṣiṇāgni*. For these three altars had to be of the same area but of different shape, the first circular, the second square and the last semicircular. Again, it has also been noted before, that

[1] "यूपावटाः पदविष्कम्भाः, त्रिपदपरिणाहानि यूपोपराणीति ।"—*BSl*, i. 112-3.

[2] *MāŚl*, i. 27.

Gārhapatya altar may have, according to certain authorities, an alternative shape of a square, besides its usual circular shape, but retaining the same area. The earliest express reference to this tradition is found in the *Śatapatha Brāhmaṇa* (c. 2000 B.C.).[1] It appears copiously in the Sūtra works.[2] From the latter we also learn of another ancient tradition that the *Dhiṣṇya*[3] may be square or circular in shape but with the same area one square *piśila*. The form of an archaic variety of the *Śmaśāna-cit* (" Fire-altar of the shape of the cemetery ") is stated to be circular, according to some peoples and square according to others. As regards its size, the *Śatapatha Brāhmaṇa*, after referring to the earlier opinions, preferably approves of an area of one square puruṣa.[4] Further instances of the early applications of the above problems are found in the *Taittirīya* and other *Saṁhitā* (c. 3000 B.C.)[5] in connexion with the construction of *Ratha-cakra-cit, Samuhya-cit, Paricāyya-cit* and *Droṇa-cit* (alternative shape). In each of these cases, one has to draw at first a square equal in area to that of the primitive *Śyena-cit*, *viz*., $7\frac{1}{2}$ square puruṣas and then that square has to be circled.[6] This has been clearly described by Baudhāyana[7] and Āpastamba[8] and it was doubtless so

[1] *ŚBr*, vii. 1. 1. 37.

[2] *BŚl*, ii. 61-3; *ĀpŚr*, xvi. 14. 1; *ĀpŚl*, vii. 5-6.

[3] *BŚl*, ii. 73; *ĀpŚr*, xvii. 21. 5; *ĀpŚl*, vii. 12-3; compare also *BŚr*, vi. 26. 29.

[4] *ŚBr*, xiii. 8. 1. 5ff.

[5] *TS*, v. 5. 4. 11. 2.

[6] In case of the *Droṇa-cit*, one-tenth of the transformed square is first deducted and the remaining rectangular portion is transformed again into a square and then circled.

[7] *BŚr*, xvii. 29; *BŚl*, iii. 183.

[8] *ĀpŚl*, xii. 12.

before.[1] We occasionally meet with other instances of this kind in the earlier Hindu works.[2]

Bürk has observed: " I shall only emphasize the fact that the Indians must have understood really in the time of the *Taittirīya Saṁhitā*, how to solve the problem of the *circling of the square* (although on very primitive methods)."[3]

[1] Compare also Bürk, *ZDMG*, LV, p. 548.

[2] *ĀpŚr*, xvi. 4. 7.

[3] *ZDMG*, LV, p. 548.

CHAPTER XII

SIMILAR FIGURES

It has been observed before that in the sacrificial rituals of the early Hindus it is oftentimes necessary to construct an altar differing in area from another by a specified amount. For instance, the *Sautrāmaṇikī-vedi* is stated to be equal to one-third of the *Mahāvedi* [1] and the *vedi* of the *Aśvamedha* double the latter.[2] The *Lakṣa-homa-vedi* and *Koṭihoma-vedi* are respectively four and twenty-five times the *Pākayājñikī-vedi.*[3] Again it is said that the primitive Fire-altar, *Caturasra-śyenacit*, should have an area of seven and a half square puruṣas at the time of the first construction. At the second construction its area shall have to be $8\frac{1}{2}$ square puruṣas; at the third $9\frac{1}{2}$ square puruṣas. In the same manner the area of the *Agni* should be increased by one square puruṣa at each successive construction up to $101\frac{1}{2}$ square puruṣas. The earliest reference to this mode of increment of the *Agni*, as has been stated before, is found in the *Śatapatha Brāhmaṇa.*[4] And the practice continued during the succeeding ages.[5] Now it is the strict injunction of the *Sruti*, that the primitive shape of the Fire-altar must not be disturbed during the course of successive constructions.[6]

[1] *BŚl*, i. 85 ; *ĀpŚl*, v. 8, 9; *KŚr*, xix. 2.2; *KŚl*, ii. 19.

[2] *ĀpŚl*, v. 10; *ĀpŚr*, xx. 9.1.

[3] *MāŚl*, ii. 6.

[4] *ŚBr*, x. 2. 3. 6 ff.

[5] *BŚl*, ii. 1 ff.; *ĀpŚr*, xvi. 17. 15. 16; *ĀpŚl*, viii. 3, 4; *KŚl*, v. 1 ff.

[6] Compare what Āpastamba remarks : "A change in the form of the *Agni* would be against the injunction of the *Śruti.*" (*ĀpŚl*, viii. 6.) According to *Śatapatha Brāhmaṇa*, those who deprive the *Agni* of its due proportions, will suffer the worse for sacrificing (x. 2. 3. 7).

Consequently in the science of the altar-construction there arose the necessity of constructing similar figures.

Now the shape of the *Mahāvedi* is that of an isosceles trapezium whose altitude is 36 *prakrama* (or *pada*), face 24 *prakrama* (or *pada*) and base 30 *prakrama* (or *pada*). Hence the *Sautrāmaṇikī-vedi* and the *vedi* of *Aśvamedha* will be of the shapes of isosceles trapeziums similar to it but in size one-third and double of it respectively. Thus we have the two following propositions:

(i) *To construct an isosceles trapezium similar to a given isosceles trapezium but with a third part of its area.*

Construct an isosceles trapezium, says Āpastamba, in the same way as the given isosceles trapezium (*Mahāvedi*) but "with $1/\sqrt{3}$ of a prakrama being substituted for a prakrama therein; or with 8 and 10 times $\sqrt{3}$ as the transverse sides and 12 times $\sqrt{3}$ as the east-west line."[1] This will be the required isosceles trapezium (*Sautrāmaṇikī-vedi*).

For the area of the constructed figure

$$= \frac{36}{\sqrt{3}} \times \frac{1}{2}\left(\frac{24}{\sqrt{3}} + \frac{30}{\sqrt{3}}\right),$$

$$= \frac{1}{3} \times 18 \times 54 = 324 \text{ square puruṣas};$$

$$\text{or} \quad = 12\sqrt{3} \times \tfrac{1}{2}(8\sqrt{3} + 10\sqrt{3}),$$
$$= 324 \text{ square puruṣas}.$$

Thus it is equal to one-third of the area of the given isosceles trapezium, which comprises 972 square puruṣas. The same construction is suggested also by Kātyāyana.[2]

[1] *ĀpŚl*, v. 8.
[2] *KŚl*, ii. 19.

(*ii*) *To construct an isosceles trapezium similar to a given isosceles trapezium but with double its area.*

The method of construction of the new isosceles trapezium will be the same, says Āpastamba, as that of the given isosceles trapezium but here "$\sqrt{2}$ of a prakrama should be taken in the place of a prakrama therein."[1] That is taught also by Baudhāyana.[2] Then the area of the new figure will be

$$= 36\sqrt{2} \times \tfrac{1}{2} (24\sqrt{2} + 30\sqrt{2}) \text{ square puruṣas},$$
$$= 1944 \text{ square puruṣas}.$$

Hence it is double the size of the given trapezium (*Mahā-vedi*).

It is thus clear that the principle underlying the early Hindu method of construction of an isosceles trapezium similar to a given one but of n times the size of it, n being integral or fractional, is practically the same as that of the given one, only the unit of measurement of the latter being replaced by another $\sqrt{n}$ times it.[3] This principle they adopted systematically for the construction of similar figures of more complicated shapes, even when the change in the size does not bear a simple relation to the size of the given figure. Thus arose the proposition:

To construct a Fire-altar similar to that of the shape of a falcon, but differing from its primitive area of $7\frac{1}{2}$ *square puruṣas by m square puruṣas.*

Baudhāyana gives the following solution of this proposition:

" Divide that which is to be the difference from the original (given) size of the altar into 15 equal parts; add

1 *ĀpŚl*, vi. 1.

2 *BŚr*, xxvi. 10.

3 Compare *BŚr*, x. 19.

to each of the (constituent) portions (*vidha*, that is, units) of the given figure two of these parts. Then construct a figure (in the same way as the given one) with $7\frac{1}{2}$ of these (altered) units."[1]

The geometrical operations to be followed in this method of construction are shortly these:[2] At first is drawn a square of an area equal to m square puruṣas. It is then divided into 15 equal parts. This may be done either by dividing one side of the square into 15 equal parts and then drawing lines parallel to the perpendicular sides, *or* by dividing one side into 3 parts and a perpendicular side into 5 parts and then drawing parallels. Two of the rectangular portions are then combined into a square and to that is again added a unit square puruṣa so as to form a third square. A side of this resulting square will be easily found to be $\sqrt{1 + 2m/15}$ puruṣas long. With this length as the unit, construct an altar in the same way as the original falcon-shaped altar. This will be the required figure. For its area will be

$$7\tfrac{1}{2} \times \left(1 + \frac{2m}{15}\right) \text{ or } \left(7\tfrac{1}{2} + m\right)$$

square puruṣas and its shape will be clearly similar to that of the given figure.

A similar method is taught briefly by Āpastamba:

"For the eight- and other-fold *Agnis*, that by which it differs from the area of the seven-fold (*Agni*), should be divided seven (and a half) ; then one part should be added to each (original) puruṣa."[3]

"For the purpose of adding parts (in puruṣa) to the (seven-fold) *Agni* together with its wings and tail, take

[1] *BŚl*, ii. 12.

[2] Compare also Thibaut's notes on the above.

[3] *ĀpŚl*, viii. 6.

for the (new) puruṣa, the (original) puruṣa increased by the seventh-fold producer of the increment. Then construct (in the same way as before)." [1]

The method has been explained more fully by Kātyāyana. His procedure is, however, slightly different from that of other writers. Kātyāyana says:

"For the purpose of adding a square puruṣa (to the original falcon-shaped *Agni*), construct a square equivalent (in area) to the original *Agni* together with its wings and tail; add to it a square of one puruṣa. Divide the sum (*i.e.* the resulting square) into fifteen parts and combine two of these parts into a square. This will be the (new) unit of square puruṣa (for the construction of the required enlarged figure)." [2]

In other words the enlarged square unit is

$$\frac{2}{15}\left(7\tfrac{1}{2} + 1\right) \quad \text{or} \quad 1 + \frac{2}{15}.$$

Or the new increased unit will be obtained thus, says Kātyāyana:

"Divide a square of one puruṣa into five parts both ways (by lines drawn cross-wise); combine five of the resulting elementary parts into a square; subtract from the sum one-third of it; add the remainder to one square puruṣa. This is another method (of determining the enlarged square unit)." [3]

[1] *Ibid*, ix. 5. Note the difference between the commentators and Bürk about the correct interpretation of these two rules of Āpastamba. I think on the whole the orthodox interpretation to be more fair and consistent with the intention of the *Śruti*. For any other interpretation would disturb the similarity of the altar at successive constructions which is expressly forbidden.

[2] *KŚl*, v. 4.

[3] *KŚl*, v. 6.

That is, the enlarged square unit is

$$1+\left(\frac{5}{5\times5}-\frac{1}{3}\cdot\frac{5}{5\times5}\right) \text{ or } 1+\frac{2}{15}$$

square puruṣas. He gives a still third method:

"Or divide a square of one puruṣa into seven parts both ways (by lines drawn cross-wise); combine seven of the (resulting elementary) parts into a rectangle, subtract from the sum (a rectangle) $1\frac{1}{7}$ aṅgulis by one puruṣa. Add the remainder to one square puruṣa. This is another method." [1]

Thus it follows that the increased square unit will be

$$1+\left(\frac{7}{7\times7}-1\times\frac{1\frac{1}{7}}{120}\right) \text{ or } 1+\frac{2}{15}$$

square puruṣas, as before.

If the number of square puruṣas (m) to be added to the original area of the *Agni* ($7\frac{1}{2}$ square puruṣas), be an exact multiple or submultiple of it, the geometrical operations are much simplified. For if $m=n\times7\frac{1}{2}$, where n may be integral as well as fractional, then the length of the new unit will be easily obtained to be equal to $\sqrt{1+n}$ puruṣa. Thus it is the $(1+n)$th *karaṇī* of a puruṣa, as has been stated by all the Śulbakāras.

This kind of increment is called *sarvābhyāsa* (or "the increment by the whole") in contradistinction to the other which is called *puruṣābhyāsa* (or "the increment by puruṣa"). [2]

[1] *KŚl*, v. 8.

[2] *KŚl*, v. 2, 3. These terms are sometimes used also in a different sense. According to it the first term signifies the increment of all the parts of the *Agni* and the second only that of the complete puruṣas in the body, wings and the tail, but not the two aratnis and the prādeśa (*ĀpŚl*, xxi. 7, 10; *KŚl*, v. 4).

The above principle of the enlargement of a *Vedi* or an *Agni* by increasing the length of the unit of measure but without altering the numbers representing the spatial magnitudes, so as to keep the form similar to the original one, is found as early as the *Śatapatha Brāhmaṇa* (c. 2000 B.C.). It says, "as large as the *Agni* (is to be made), so large (should be made) its units of measure; and by so much one measures it in the same way (as before)."[1] To construct a *Vedi* 14 or $14\frac{3}{7}$ times as large as the *Mahāvedi*, and which will be similar to it, this *Brāhmaṇa* says:

"As much this *vedi* of the seven-fold Fire-altar is, making fourteen times, so much, one measures the *vedi* of 101-fold. Or then he measures (by means of) a cord 36 prakramas long; folds it into 7 equal parts; of these three parts he adds to the east-west line and throws out 4. Then he measures 30 prakramas; folds them into 7 equal parts; of these three parts he adds to the hind (transverse line) and throws out 4. Then he measures 24 prakramas; folds them into 7 parts; of these 3 parts he adds to the front (transverse line) and throws off 4. This, then, is the alternative measurement of the (enlarged) *Vedi*."[2]

Here two methods are to be discerned. According to one, it is required to construct a *vedi* 14 times as much while according to the other $14\frac{3}{7}$ times as much.[3] The operations implied are doubtless as follows:[4] The altar-

[1] *ŚBr*, x. 2. 1. 3, 11.

[2] *Ibid*, x. 2. 3. 7-10.

[3] That is, the *Mahāvedi* on which the primitive Fire-altar is raised, is enlarged in proportion to the size of the latter. Now there are two opinions (*vide infra*) as regards the method of construction of the maximum Fire-altar. According to one the latter becomes 14 times the *Sapta-vidha Agni*, while according to the other $14\frac{3}{7}$ times. The *Mahāvedi* is enlarged in the two schools accordingly.

[4] Eggeling's explanation (*SBE*, Vol. XLIII, pp. 310-311, footnotes) is obviously erroneous. That will make the *vedi* too great.

builder should at first construct a square of sides 36 prakramas each. Fourteen such squares should be combined into a large square. A side of the resulting square, which is clearly equal to $36\sqrt{14}$ prakramas, is taken for the east-west line. Or otherwise a smaller square should be divided into the 7 equal rectangular portions by drawing lines parallel to a side of the square; the rectangle comprising three of these strips should be transformed into a square and then combined with the former larger square. A side of the resulting square, which will be easily recognised to be equal to $36\sqrt{14\frac{3}{7}}$ prakramas, is taken for the east-west line of the enlarged *vedi* to be constructed. By the same kind of operations, the face of the new *vedi* is obtained to be $24\sqrt{14\frac{3}{7}}$ prakramas and its base $30\sqrt{14\frac{3}{7}}$. So that the size of the new *vedi* will be $14\frac{3}{7}$ times that of the *Mahāvedi* and its shape similar to that of the latter.

The following method has been taught for the construction of a falcon-shaped Fire-altar. 14 or $14\frac{3}{7}$ times as large as the primitive one and similar to it.

" Now the construction of (the enlarged forms of) the *Agni*: Twenty-eight (square) puruṣas are in front and twenty-eight (square) puruṣas behind; this is the body (of the *Agni*). Fourteen (square) puruṣas form the southern wing; fourteen the northern wing and fourteen the tail. Fourteen *aratnis* (meaning, a length equal to the side of a square comprising fourteen square aratnis (*i.e.*, $\sqrt{14}$ aratnis) is added to the southern wing; fourteen aratnis to the northern wing; and fourteen vitastis (meaning, a length equal to the side of a square of fourteen square vitastis, *i.e.*, $\sqrt{14}$ vitastis) to the tail. This is the measurement of (the *Agni* of) 98 square puruṣas with a little excess (due to the increment of the wings and the tail). Or again he measures (by means of) a cord three puruṣas; folds it into seven (equal) parts; of them four parts he adds to the body and three to the wings and tail. Then he measures an aratni; folds it into seven parts; of them three parts he adds to the southern wing, three to the northern wing and throws off one.[1] Then he measures a vitasti; folds it into seven parts; of them three parts he adds to the tail and throws out four. Thus is constructed the 101-fold (*Agni*) and it corresponds with the *vedi* of this."[2]

The geometrical operations intended to be performed in this case are similar to those indicated in the previous case. Here also are to be discerned two methods for the construction of the *Ekaśata-vidha Agni* or the Fire-altar of $101\frac{1}{2}$ square puruṣas. According to the first method each unit of measure is $\sqrt{14}$ times the unit of measure employed in the construction of the primitive Fire-altar of $7\frac{1}{2}$ square puruṣas, while according to the other, probably the more ancient one, it will be $\sqrt{14\frac{3}{7}}$ times. Though the principle of similarity of shapes is perfectly maintained in either methods, as regards the size, they of course yield only approximate results. For $7\frac{1}{2} \times 14 = 105$ square puruṣas ; $7\frac{1}{2} \times 14\frac{3}{7} = 108\frac{3}{14}$ square puruṣas. The *rationale* of these results we shall indicate in the next chapter. We should but note here that as the second method[3] yields a result more deviating from the correct and desired one, *viz.*, $101\frac{1}{2}$ square puruṣas, its

[1] The printed text has *catura* meaning "four." It is obviously wrong. So I have amended it to "one."

[2] *ŚBr*, x. 2. 3. 11-14. Eggeling's rendering is erroneous.

[3] The second method seems to have been meant in the beginning for an enlargement on the second plan, to be explained shortly. So that it is only by mistake that it was employed for the enlargement on the first plan.

correctness was challenged even before the time of the *Śatapatha Brāhmaṇa* (c. 2000 B.C.).

"As to this they say, 'as seven[1] (square) puruṣas have exceeded, how is it that they do not deviate from the right total (*sampad*, which the altar ought to have).'"

The change of the representative number is strictly forbidden in this work, as that will disturb the principle of similarity of the forms. It observes:

"Now some intending to construct subsequent (larger) forms (of the *Agni*), increase (the number of) these prakramas and vyāmas, supposing, 'we shall enlarge the womb (*yoni*) accordingly.' One should not do so; for the womb does not increase along with the child that has been born; but indeed only as long as the child is within the womb, so long does the womb enlarge; and this much again is the growth of the child here. Those who do it in that way, certainly do they deprive this Father Prajāpati of his perfection (*sampad*; that is, due proportions)." [2]

In the foregoing methods of enlargement of the original size of an altar, it will be observed, all the constituent parts of it receive increments in equal proportions. But sometimes an altar, particularly the falcon-shaped one, is enlarged on an entirely different plan. According to

[1] The printed text has *trayodaśa* or "thirteen." It seems to be a mistake. For it was intended to construct a Fire-altar having an area equal to $101\frac{1}{2}$ square puruṣas. Now the method adopted produces one with an area of $108\frac{3}{14}$ square puruṣas. So $108\frac{3}{14} - 101\frac{1}{2}$ or $6\frac{5}{7}$ square puruṣas are in excess. In round numbers this might be stated as equal to 7 square puruṣas. But "thirteen" cannot be justified. So I think that here again the text should be amended to *sapta* or "seven."

[2] *ŚBr*, x. 2.3. 6-7. In this metaphor, the *Agni*, as usual, is called the Father Prajāpati; his child is the unit of measure employed; and the womb is his form and hence the numbers representing the spatial relations of that form (*cf.* *ŚBr*, viii. 6.3.12; x. 2.3.3). This passage has been referred to in the aphorism vi. 4 of the *Kātyāyana Śulba*.

it, certain parts of the altar, such as the complete puruṣas in the body, wings and tail of the falcon, are enlarged proportionately, while the rest, the two aratnis and the prādeśa by which the wings and tail were lengthened fundamentally, are left unaffected. This plan of enlargement of the falcon-shaped altar has been noticed by all the three principal Śulbakāras. It is particularly advocated by the tradition of one school as necessary for the altar of the Horse Sacrifice. Other schools, however, follow the general plan of proportionate increment of all parts in that case too.[1]

We have so far considered the construction of figures, particularly isosceles trapeziums, similar to a given one and differing from it in area by a specified amount. It is sometimes also necessary to construct a figure, similar to another and having a given side. For instance, the *Ekādaśinī-vedi* is stated to be similar to the *Mahāvedi* and is indeed an enlarged form of it.[2] It is socalled because it must have, according to the injunction of the scriptures, eleven (*ekādaśa*) sacrificial posts (*yūpa*) in front. It is further prescribed that the two posts on either sides of the middle one must be at a distance of one *akṣa* (=104 aṅgulis) and four aṅgulis from it and the rest are an akṣa distant from each other. Each post has a diameter of one pada. Hence the east side of the *Ekādaśinī-vedi* is 10 akṣas 11 padas 8 aṅgulis long. Thus it is required to construct an isosceles triangle similar to the *Mahāvedi* and having its face equal to this length. Again, according to Baudhāyana, the shape of the *Aśvamedha-vedi* is similar to the *Mahāvedi* and has 21 *yūpas* on the east side, that is, has a face of 20 akṣas 21 padas 8 aṅgulis long.

[1] *BSl*, ii. 8 ff.; iii. 321-3; *ĀpSl*, xxi. 6-10; *KSl*, v. 7.
[2] *BSl*, i. 106 ff.

The method of construction to be adopted in these cases is indicated by Baudhāyana thus:

"For that (*Ekādaśinī-vedi*) take the twenty-fourth part of 10 akṣas 11 padas 8 aṅgulis; this will be the prakrama and with this (altered unit), the *vedi* has to be constructed (in the same way as the *Mahāvedi*)."[1]

"For the *Aśvamedha-vedi* take the twenty-fourth part of 20 akṣas 21 padas 8 aṅgulis; this will be the prakrama and with this (modified unit), the *vedi* has to be constructed (in the same way as the *Mahāvedi*)."[2]

The face, base and altitude of the *Mahāvedi* are given respectively to be 24, 30 and 36 prakramas in length. If a, b, c be the corresponding quantities of a similar figure, we shall have

$$\frac{a}{24} = \frac{b}{30} = \frac{c}{36};$$

$$b = 30\left(\frac{a}{24}\right), \quad c = 36\left(\frac{a}{24}\right).$$

This is equivalent to the change of the ordinary unit to $a/24$ times it, where a is the given length of the face of the figure to be constructed.

A similar method will have to be followed in constructing an isosceles trapezium similar to a given one and having a given altitude.[3]

These methods are also taught by Kātyāyana.[4]

It may be noted that Āpastamba's method of construction of several altars by employing the rational rectangle (15, 36, 39) by successively varying the units, is obviously equivalent to the construction of a system

[1] *BŚl*, i. 107; *vide* also *BŚr*, xxvi. 23.

[2] *BŚl*, i. 108; *vide* also *BŚr*, xxvi. 10.

[3] *BŚl*, i. 109-10.

[4] *KŚl*, vii. 1-3; compare *KŚr*, viii. 8.22.

of rectangles similar to it. Indeed the science of the altar-construction requires the drawing of (a system of) similar rectangles, triangles, rhomboids, circles and other figures. For the enlargement of altars of those shapes involves it.

CHAPTER XIII

Geometrical Algebra

The geometrical constructions described in the preceding chapter are of considerable algebraic significance. They indeed form the seed of the Hindu *geometrical algebra* whose developed form and influence we notice as late as in the *Bījagaṇita* of Bhāskara II (born 1114 A.D.). The first plan of enlargement of a figure in which all the constituent parts are affected in equal proportions, leads to the quadratic equation of the type

$$ax^2 = c.$$

The second plan leads to the complete quadratic equation

$$ax^2 + bx = c.$$

Let x denote the length of the enlarged unit of puruṣa and m denote the total increment in area. Then, in the case of the enlargement of the *isosceles trapezium* on the first plan, we shall have

$$36x \times \left(\frac{24x + 30x}{2} \right) = 36 \times \left(\frac{24 + 30}{2} \right) + m,$$

$$\text{or} \quad 972x^2 = 972 + m.$$

Therefore $$x^2 = 1 + \frac{m}{972}$$

Hence $$x = \sqrt{1 + m/972}\,\cdot$$

If $m = 972(n-1)$, so that the area of the enlarged trapezium is n times its original area, we get

$$x = \sqrt{n}$$

as given in the *Sulba*. The particular cases, when $n =$

14 or $14\frac{3}{7}$, are found as early as the *Śatapatha Brāhmaṇa* (*c.* 2000 B.C.).[1]

For the enlargement of the *falcon-shaped Fire-altar* on the *first plan*, we get the quadratic equation

$$2x \times 2x + 2\left\{x \times \left(x + \frac{x}{5}\right)\right\} + x \times \left(x + \frac{x}{10}\right) = 7\tfrac{1}{2} + m,$$

or
$$\frac{15}{2}x^2 = 7\frac{1}{2} + m.$$

Therefore
$$x^2 = 1 + \frac{2m}{15}.$$

Hence
$$x = \sqrt{1 + \frac{2m}{15}}.$$

In particular, when $m=94$, that is, when the Fire-alter has its maximum enlargement permissible under the *Śulba*, we have

$$x^2 = 13\frac{8}{15} = 14, \text{ approximately}$$

as found in the *Śatapatha Brāhmaṇa* (*vide supra*).

In the case of the enlargement of the falcon-shaped Fire-altar on the *second plan*, the geometrical operations are equivalent to the solution of the following complete quadratic equation:

$$2x \times 2x + 2\{x \times (x + \tfrac{1}{5})\} + x \times (x + \tfrac{1}{10}) = 7\tfrac{1}{2} + m,$$

$$\text{or} \quad 7x^2 + \tfrac{1}{2}x = 7\tfrac{1}{2} + m.$$

Completing the square on the left hand side, we get

$$(7x + \tfrac{1}{4})^2 = \frac{841}{16} + 7m$$

[1] *ŚBr*, x. 2. 3. 7 ff. *Vide supra* pp. 158 f.

Therefore $$x = \frac{1}{28}(\sqrt{841+112m} - 1), \quad \ldots \quad (1)$$

$$= \frac{1}{28}\left\{29\left(1 + \frac{56m}{841}\right) - 1\right\},$$

$$= 1 + \frac{2m}{29},$$

neglecting higher powers of m. Therefore we have

$$x^2 = 1 + \frac{4m}{29}, \text{ approximately,} \quad \ldots \quad (2)$$

Kātyāyana says:

"Or for the second and following constructions, increase (the usual unit of) the (square) prakrama by itself for every seven constructions; so that (at each successive construction) take for the prakrama, the original value of the prakrama enlarged by its one seventh."[1]

So that, according to him, the enlarged unit (x^2) will be

$$x^2 = 1 + \frac{m}{7}; \quad \ldots \quad (3)$$

whereas a more accurate value has been proved above to be

$$x^2 = 1 + \frac{m}{7\frac{1}{4}}, \text{ approximately,} \quad \ldots \quad (2.1)$$

In particular, when the Fire-alter has its maximum enlarged form of $101\frac{1}{2}$ square puruṣas, we have

$$7\tfrac{1}{2} + m = 101\tfrac{1}{2}$$

Substituting in (1), we get

$$x = \frac{1}{28}(\sqrt{11369} - 1).$$

[1] *KŚl*, vi. 3.

Hence $$x^2 = \frac{1}{784}(11370 - 2\sqrt{11369}).$$

Now $$\sqrt{11369} = 106 + \frac{133}{212}, \text{ approximately,}$$

Therefore $$x^2 = \frac{1}{784}\left(11156 + \frac{79}{106}\right),$$

$$= 14 + \frac{19159}{83104},$$

$$= 14 + \cfrac{3}{13\frac{245}{19159}}.$$

Hence $$x^2 = 14 + \frac{3}{13}, \text{ approximately} \qquad \dots (4)$$

Kātyāyana gives a nearly equal value,

$$x^2 = 14 + \frac{3}{7} \qquad \dots (5)$$

He says:

"The side which turns out a square of the area fourteen and three-sevenths (square) prakramas will be the length of the prakrama for the 101-fold (Fire-altar)." [1]

It is not at all easy to determine the *rationale* of Kātyāyana's formulæ (3) and (5), whether they were obtained in the way indicated above by the method of the solution of a complete quadratic equation, or in any other way. It is found that there are discrepancies between the results calculated by the former method and those found in the extant copies of the manuscripts of his works. How to explain them? Whether by the crudeness of his method of solving the quadratic equation or of the

[1] *KŚl*, vi. 2.

calculations attendent on it,[1] if he had at all followed that method, or in any other way? Now the discrepancy need not be considered very serious in the general case, especially if we remember the degree of accuracy that can be naturally expected in those early times, or that is ordinarily found to have been followed then. Even in modern times 7 will be considered to be a very fair approximation to $7\frac{1}{4}$. What will appear to be serious is in the other case which requires an emendation of the existing text in order to explain away the discrepancy with the result calculated as above. But, we would remark, that by itself should not be considered a very formidable objection against our hypothesis as regards the method of Kātyāyana. For those who have dealt with ancient manuscripts are quite aware that they doubtless require emendations here and there. That discrepancy can be explained away much more easily and reasonably by supposing that the result (5) was derived from the modified result (3) by substituting the value, $m = 94$. but was not calculated directly from the equation (1) as we have done above.

It should be noted that the relevant portions of the existing manuscripts of the *Kātyāyana Śulba* might be considered to be quite correct and his results extremely accurate if we follow a certain interpretation of the text. Let us suppose that (1) the term *ekaśata-vidha* means " the construction (of a Fire-altar having an area) of 101

[1] It should be noted that the result (4) does not follow, quite contrary to expectations, from the equation (2), on the substitution in the latter of the value, $m = 94$.

$$x^2 = 1 + \frac{4 \times 94}{29} = 13\frac{28}{29} = 14, \text{ approximately.}$$

This is too less than the value (5) recorded by Kātyāyana. That is why we started by substituting $m = 94$ in the equation (1). This shows that the process of calculation has much to account for.

(square puruṣas)," and (2) that the area of 101 square puruṣas comprises only the complete puruṣas in the altar, not including the two aratnis and the prādeśa. Then the algebraic representation of the second method of enlargement of the primitive falcon-shaped Fire-altar will be

$$7x^2 = 7 + m.$$

Therefore $$x^2 = 1 + \frac{m}{7}.$$

And in the particular case of the maximum enlargement, we shall have

$$7x^2 = 101.$$

Hence $$x^2 = 14\frac{3}{7}.$$

Thus the results come out exactly to be the same as are found in extant MSS. According to this interpretation, the construction of the enlarged altars does not even involve the solution of the complete quadratic equation.

We shall now proceed to examine how far the above interpretation can be held to be correct. It should be opposed mainly on three grounds: *Firstly*, the supposed interpretation of the term *ekaśata-vidha* as meaning the construction of a Fire-altar of an area of 101 square puruṣas is very unusual. Indeed, according to all the *Śulba* and also the *Śatapatha Brāhmaṇa* that term must always refer to an area of 101½ square puruṣas. Similarly, it is known, that *sapta-vidha* always refers to 7½ square puruṣas, *aṣṭa-vidha* to 8½ square puruṣas and so on. *Secondly*, for that interpretation the stipulated area of 101 square puruṣas has been assumed to be comprised of the area of the complete puruṣas in the body, wings and tail of the falcon-shaped Fire-altar, in exclusion of the two aratnis and the prādeśa in the latter. Therefore in that case the area of the complete altar, all told, will be greater than 101½ square puruṣas. But the maximum extent up to which the Fire-altar is sanctioned to be enlarged by successive constructions is 101½ square puruṣas. Thus Baudhāyana expressly remarks:

" The first (*i.e.*, when constructed for the first time) Fire-altar has an area of seven and a half square puruṣas. The second contains eight and a half; the third, nine and a half. In this way, the addition of one puruṣa takes place at each successive construction up to the *ekaśata-vidha* (*i.e.*, ' the construction including an area of 101½ square puruṣas '). After that, the *ekaśata-vidha Agni* should be repeated (without making any further enlargement of it.) Or else the sacrifice should be performed without an *Agni*."[1]

Śatapatha Brāhmaṇa says:

" Some say, '*eka-vidha Agni* should be constructed first; then by an increment of one (square puruṣa) successively up to a construction of unlimited size.' One should not do so. The *Prajāpati* (*i.e.*, *Agni*) was created first as *sapta-vidha* indeed. Proceeding to reconstruct himself, he stopped at the *ekaśata-vidha* ('a construction comprising 101½ square puruṣas'). He who constructs (a Fire-altar) smaller than the *sapta-vidha*, cuts asunder this *Prajāpati*: he voluntarily becomes a sinner as one would be by destroying or injuring his better. Again one who constructs (a Fire-altar) exceeding the *ekaśata-vidha*, he proceeds beyond all these (visible Universe), for the *Prajāpati* is this Universe. Hence one should first construct the *sapta-vidha* and then by the increment of one (square puruṣa) in succession up to the *ekaśata-vidha*. But one should not construct in excess of the *ekaśata-vidha*. Thus he

[1] *BŚl*, ii. 1-7. Compare also *ĀpŚl*, viii. 3; *KŚr*, xvi. 8.25; *KŚl*, v. 1; vi. 4; *ĀpŚr*, xvi. 17. 15-6. In this last work Āpastamba has expressly referred to the tradition of the Vājasaneya school (*i.e.*, to the *Śatapātha Brāhmaṇa*) on this point (*vide* below).

neither cuts asunder the Father *Prajāpati*, nor does he proceed beyond all these (the Universe)." [1]

So the above interpretation stands in direct contradiction with the injunctions of the early scriptures.

Thirdly, the practice of the mention of an altar by reference to the area of a part of it, as has been supposed for the above interpretation, is unknown in the *Śulba* and earlier literatures of the Hindus. Thibaut once thought that such a thing had been implied in a certain rule of Baudhāyana. He says, "But according to the above sūtra[2] and its commentary the *Āśvamedhika Agni* was of a different nature. It had to comprise twenty-one puruṣas not including the lengthening of the two wings by one aratni each and of the tail by one prādeśa, so that its ātman consisted of twelve square puruṣas, its wings and tail of three puruṣas each. A proportional increase of the two aratnis and the prādeśa would amount to $1\frac{1}{2}$ square puruṣas and then the agni would no longer be 'ekaviṃśa' as the śruti demands. Therefore the wings were lengthened only by the regular aratni (of 24 aṅgulis) and the tail by the regular prādeśa (12 aṅgulis), so that the increase of the agni caused thereby remained less than one square puruṣa and the agni preserved its character of ekaviṃśa." [3] Thibaut has been followed in this matter by Bürk.[4] They are clearly in error.[5] For the sūtra they had in view says absolutely nothing about the actual size of the *Āśvamedhika Agni*. The commentator Dvārakanātha

[1] *ŚBr*, x. 2.3. 17-8.

[2] The reference is to the following sūtra of the *Baudhāyana Śulba* (iii. 321) : "आश्वमेधिकस्याग्नेः पुरुषाभ्यासो नारत्निप्रादेशानाम्।"

[3] *The Paṇḍit*, New Series, Vol. I, p. 769.

[4] *ZDMG*, LVI, pp. 355 ff.

[5] Eggeling was wrong in supposing that *ekaviṃśa Agni* means "an altar measuring twenty-one man's length on each of the four sides of its body." (*ŚBr*, V, p. 334, fn. 2.)

Yajvā truly took the area of the Fire-altar to be $21\frac{1}{2}$ square puruṣas. His use of the *trikaraṇī* of a puruṣa in the construction can be more easily and satisfactorily explained in a way different from that supposed by Thibaut and Bürk.

Now as regards the size of the *Āśvamedhika Agni*, Kātyāyana states clearly: "(It) is twice or thrice (the area of) the primitive *Agni* or twenty-onefold (*ekaviṃśa-vidha*)."[1] Such is also the opinion of the other Śulbakāras as well as of the *Brāhmaṇa*.[2] *Śatapatha Brāhmaṇa* mentions also of a particular school, according to which the total area of the altar should be $12\frac{1}{2}$ square puruṣas.[3] As regards the method of enlargement of the primitive falcon-shaped altar of $7\frac{1}{2}$ square puruṣas by which we are to arrive at the construction of the altar for the Horse Sacrifice having the specific area sanctioned for it by the scriptures, one school advocates the proportional enlargement of every part of the construction, while another school the similar enlargement of only those parts which comprise the complete puruṣas, excluding the two aratnis in the wings and the prādeśa in the tail. For an altar with an area of 15 or $22\frac{1}{2}$ square puruṣas enlarged on the first plan we shall have, it will be easily found, to take the *dvikaraṇī* or *trikaraṇī* of a puruṣa and proceed in the same way as in the construction of the primitive *Agni*. For an altar with an area of $21\frac{1}{2}$ square puruṣas enlarged, but according to the second plan, the algebraic equation will be

$$7x^2 + \tfrac{1}{2}x = 21\tfrac{1}{2}.$$

Therefore $x = \frac{1}{28}(\sqrt{2409} - 1)$.

[1] *KŚr*, xx. 4.15; *KŚl*, v. 2-3.

[2] *BŚl*, iii. 8 ff. ; iii. 321 ff. ; *ĀpŚl*, xxi. 6, 9; *ĀpŚr*, xx. 9.1; *ŚBr*, xiii. 3-3. 7 ff.

[3] *ŚBr*, xiii. 3.3.9.

Hence $$x^2 = \frac{1}{784}(2410 - 2\sqrt{2409})$$

$$= \frac{1}{784}\left(2410 - 2 \times 49\frac{4}{49}\right), \text{ approximately}$$

$$= 3 - \frac{123}{2401}.$$

Therefore we may take as a very near approximation $x^2 = 3$. And that is what Dvārakanātha Yajvā has done.[1] We do not think it to be fair to him to interpret him otherwise as Thibaut does.

For the above reasons we discard this latter interpretation of Kātyāyana's rules and hold that the construction of altars enlarged according to the second plan noted above and which is used to be followed in a certain school of the *Brāhmaṇa*, does undoubtedly depend preliminarily on the solution of the complete quadratic equation,

$$ax^2 + bx = c,$$

and further that Kātyāyana's results were obtained in this way. Milhaud is stated to have arrived at the former conclusion before.[2] His article is unfortunately not available to me and so I do not know the arguments adduced by him.

It may be noted that the commentator Mahīdhara is of no help to us about this knotty point of the enlargement of the Fire-altar. According to his interpretation of

1 We obtain nearly the same result from the equation (2). For putting $m = 14$ ($= 21\frac{1}{2} - 7\frac{1}{2}$) in it, we get

$$x^2 = 3 - \frac{2}{29} = 3, \text{ nearly}.$$

2 G. Milhaud, "La Géométrie d'Apastamba," *Revue génerate des Sciences*, XXI, 1910, pp. 512-520; quoted by Professor D. E. Smith in his *History of Mathematics*, Vol. II, p. 444 fn.

Kātyāyana's rules, the area of the Fire-altar after an increment of m square puruṣas will be $7\frac{1}{2} \times \left(1 + \frac{m}{7}\right)$ square puruṣas. This interpretation evidently assumes a proportional enlargement of all the constituent parts of the primitive Fire-altar and hence should be discarded as being directly against the express intention of the *Śulba*. Another objection against it will be that the area of the Fire-altar at the final construction (the ninety-fifth construction) according to it will be $7\frac{1}{2} \times \left(1 + \frac{94}{7}\right)$or $108\frac{3}{14}$ square puruṣas and hence much in excess of the maximum enlarged area, *viz.*, $101\frac{1}{2}$ square puruṣas, permissible under the *Śulba*. Weber [1] informs us that such a method of enlargement of the Fire-altar is found in the *Paddhati* of Yājñikadeva. Its origin seems to be in the *Śatapatha Brāhmaṇa*.[2] But it was criticised and challenged even then.[3] I cannot give the opinion of the commentator Rāma on this point as the relevant portion of his commentary is not available to me at present.

The solution of the quadratic equation in its simpler forms is required in connexion with the enlargement of altars followed in a different school According to that school, we are informed,[4] the altars with areas $1\frac{1}{2}$, $2\frac{1}{2}$,... $6\frac{1}{2}$ square pusuṣas should be of the square form. So it will then be necessary to construct a square differing from another by a specified quantity. Now we find in the

1 A. Weber, *Indische Studien*, XIII, p. 240 f.

2 *ŚBr*, x. 2. 3. 11-14.

3 *ŚBr*, x. 2. 3. 15-6.

4 *BŚl*, iii. 319.

Śulba, the following general rule for the enlargement of a square:

"Add on the two sides (of the given square), those (two rectangles) which are described with as much as the increment (of the side of the square) and its side; add further at the corner, the square which is produced by that increment." [1]

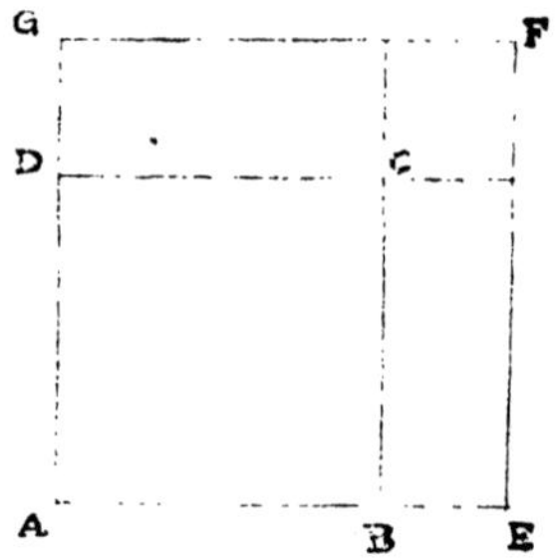

Fig. 74

Let $ABCD$ be the given square. Suppose its side is to be increased by the amount BE, say. Add to the sides BC and DC two rectangles CE and CG each of which is equal to $AB.BE$. Also add at the corner C, the square CF equal to the square on the increment BE of the side AB. Then $AEFG$ is the square on the enlarged side AE.

This is the analogue of the algebraic identity,

$$(a+b)^2 = a^2+2ab+b^2.$$

Now we shall apply this result to enlarge a square of area a^2 by m square puruṣas, say. If x be the increment of a side, then by the above rule

$$x^2 + 2ax = m,$$

$$\text{or} \quad x^2 + 2ax + a^2 = a^2 + m.$$

[1] *ĀpŚl*, iii. 9. Compare also *BŚl*, iii. 192-4.

Therefore $x = \sqrt{a^2 + m} - a.$

The geometrical method of the transformation of a square (c^2) into a rectangle having a given side (a), which has been described before, is obviously equivalent to the solution of the linear algebraic equation,

$$ax = c^2.$$

CHAPTER XIV

INDETERMINATE PROBLEMS

In the *Śulba* we find the solution of certain very interesting indeterminate problems. Some of them follow directly from the geometrical constructions made therein; others have been tacitly assumed for that purpose. We shall here notice the more important ones amongst them.

Rational Right-angled Triangles.

The formula of Kātyāyana to find the sum of n equal squares of sides a each amounts to

$$a^2(\sqrt{n})^2 + a^2\left(\frac{n-1}{2}\right)^2 = a^2\left(\frac{n+1}{2}\right)^2.$$

Putting m^2 for n, in order to make the sides of the right-angled triangle free from the radical, and dividing out by a^2 we get

$$m^2 + \left(\frac{m^2-1}{2}\right)^2 = \left(\frac{m^2+1}{2}\right)^2. \qquad \text{(I)}$$

as the solution of the indeterminate equation of the second degree

$$x^2 + y^2 = z^2. \qquad \text{(A)}$$

If the sides of the right-angled triangle are to be integral as well as rational, m must be odd. According to Proclus (c. 450 A.D.), this solution was known to Pythagoras (c. 540 B.C.).

A still more general solution of (A) which includes the solution (I) as a particular case, is furnished by the method of transformation of a rectangle into a square. It is this:

$$\left(\sqrt{mn}\right)^2 + \left(\frac{m-n}{2}\right)^2 = \left(\frac{m+n}{2}\right)^2,$$

where m, n are any two arbitrary numbers. Substituting p^2, q^2 for m, n respectively, in order to eliminate the irrational quantities, we get

$$p^2q^2 + \left(\frac{p^2-q^2}{2}\right)^2 = \left(\frac{p^2+q^2}{2}\right)^2. \qquad \text{(II)}$$

A further generalisation has been tacitly assumed in some methods employed by Āpastamba for the construction of the *Mahāvedi*. If the sides of a rational right triangle are known, then other rational right triangles can be obtained by multiplying them by any rational integer, or how he puts it, by increasing them by any rational multiples of them. Thus if α, β, γ be a rational solution of $x^2+y^2=z^2$, then other rational solutions of it will be given by $l\alpha$, $l\beta$, $l\gamma$ where l is any rational number. This is clearly in evidence in the formula of Kātyāyana. It is now known that all rational solutions of (A) can be obtained without duplication in this way.

Karavindasvāmī gives the solution,[1]

$$x, \left(\frac{m^2+2m}{2m+2}\right)x, \left(\frac{m^2+2m+2}{2m+2}\right)x.$$

It can of course be derived from the solution (I) by multiplying by x and dividing by m^2 every element of it and then putting $m+1$ for m.

Rational Right Triangles having a Given Leg.

There seems to have been an attempt to find the rational right triangles having a given leg. We find in

[1] *Vide* his commentary on *ĀpŚl*, i. 4.

the *Śulba* at least two such right triangles having a common leg a, *viz.*,

$$(a,\ 3a/4,\ 5a/4) \text{ and } (a,\ 5a/12,\ 13a/12).$$

The principle underlying these solutions will be easily detected to be that of the reduction of the sides of any rational right triangle in the ratio of the given leg to the corresponding side of it. Thus the sides of a rational right triangle having a given leg a will be $(a,\ a\beta/\alpha,\ a\gamma/\alpha)$, where $\alpha,\ \beta,\ \gamma$ are the sides of any rational right triangle. So that starting with the solution (II), we shall find that all rational right triangles having the leg a will be given by

$$a,\ \left(\frac{p^2-q^2}{2pq}\right)a,\ \left(\frac{p^2+q^2}{2pq}\right)a.$$

But this general solution is not stated anywhere in the *Śulba*. The commentators of Āpastamba, however, give the solution

$$a,\ \left(\frac{m^2+2m}{2m+2}\right)a,\ \left(\frac{m^2+2m+2}{2m+2}\right)a.$$

It should be noted that the above principle for obtaining the rational right triangles having a given leg has been followed expressly in later times in India by Mahāvīra (850) and in Europe by Leonardo Fibonacci of Pisa (1202) and Vieta (*c.* 1580).

Simultaneous Indeterminate Equations.

The construction of the *vedi* gives rise to a type of problems of indeterminate character, though in a few cases the physical conditions are so prescribed as to make them determinate.[1] For instance, the breadth of the *Gārhapatya*

[1] See Bibhutibhusan Datta, "The Origin of Hindu Indeterminate Analysis," *Archeion*, XII (1931), pp. 401-407.

vedi shall have to be, according to the tradition of the scriptures, one vyāyāma.[1] But its form should be square according to some and circular according to others.[2] It is to be constructed with five layers of bricks, each layer consisting of 21 bricks. Further the rifts of bricks in two consecutive layers must not coincide. The shape of the bricks employed may be square or rectangular. Now the most interesting case is that in which the *vedi* and also the bricks employed in constructing it, are square in shape. It is said that three kinds of square bricks should be made with the sixth, fourth and third part of a vyāyāma as a side. The first layer (l_1) should be constructed with 9 bricks of the first kind and 12 bricks of the second kind, and the second layer (l_2) with 5 bricks of the third kind and 16 bricks of the first kind.[3] That is, if we denote the bricks of different kinds thus: $b_1 = v^2/36$, $b_2 = v^2/16$, $b_3 = v^2/9$, where v denotes a vyāyāma, then

$$l_1 = 9b_1 + 12b_2, \quad l_2 = 5b_3 + 16b_1.$$

The third and fifth layers are replica of the first and the fourth, of the second.

Now it may be asked how the ancient altar-builders determined the size of the bricks of different kinds and the number of bricks of each kind that will be required for the construction of each layer. They proceeded probably in some way like this: Since 21 is not a square number, no layer of the altar can be constructed with bricks of the same kind. So the number of kinds of bricks employed in any layer must be at least 2. Since no two successive layers should have identical cleavage, all the bricks

[1] *BŚl*, ii. 61; *ĀpŚl*, vii. 5; compare also *ĀpŚl*, vii. 10.

[2] *BŚl*, ii. 62-3; *ĀpŚl*, vii. 6.

[3] *BŚl*, ii. 66-9.

employed in the second layer must not be the same as those of the first layer. Hence the minimum kinds of bricks must be 3. Assume then their sides to be p, q and rth part of a vyāyāma, where p, q, r are rational integers to be determined. Suppose the first layer consists of x bricks of the first kind and y bricks of the second kind. Then we must have

$$\frac{x}{p^2} + \frac{y}{q^2} = 1,$$

$$x + y = 21.$$

Similarly if the second layer consists of u bricks of the third kind and v bricks of the first kind,

$$\frac{u}{r^2} + \frac{v}{p^2} = 1$$

$$u + v = 21.$$

Thus we are led to the simultaneous indeterminate equations

$$\left.\begin{aligned} \frac{x}{m^2} + \frac{y}{n^2} &= 1 \\ x + y &= 21. \end{aligned}\right\} \quad \ldots \quad \text{(A)}$$

Baudhāyana's statements about the size of the different varieties of bricks and the number of them employed in the construction of a particular layer, amount to the following solutions of the equations (A)

$$\left.\begin{aligned} m = 6,\ x = 9, \\ n = 4,\ y = 12; \end{aligned}\right\} \quad \left.\begin{aligned} m = 3,\ x = 5, \\ n = 6,\ y = 16. \end{aligned}\right\}$$

These solutions were probably obtained by trial in succession, thus: Solving (A) as simultaneous linear equations in x and y, we get

$$x = \frac{m^2(21 - n^2)}{m^2 - n^2},$$

$$y = \frac{n^2(m^2 - 21)}{m^2 - n^2}.$$

Now the physical circumstances of the problem are such that x and y must be positive integers. Therefore if $m > n$

$$m^2 > 21 > n^2$$

or

$$m > \sqrt{21} > n.$$

Since

$$5 > \sqrt{21} > 4$$

therefore we must have

$$m \geq 5, n \leq 4.$$

If on the contrary $m < n$, we must have

$$n^2 > 21 > m^2$$

Hence

$$n \geq 5, m \leq 4.$$

Then substituting in the expression for x one after another the values $n = 4, 3, 2, 1$, we can determine by trial the value of m which will make the value of x in each case integral. Thus we shall easily arrive at the solutions given by Baudhāyana.

A much more difficult problem of the same type arises in connexion with the construction of the falcon-shaped Fire-altar. Its total area is given to be $7\frac{1}{2}\, a^2$, where a=a puruṣa. It is laid down that the altar must be constructed in fire layers and the number of bricks employed in any layer must be 200. As before, the rifts of bricks in any two successive layers must not coincide. There is no injunction of the scriptures about the varieties of the bricks to be employed or about their relative size and shape. In one method of construction Baudhāyana employs four kinds of square bricks whose sides are respectively the fourth, fifth, sixth and tenth part of a. Let us take, in general, the areas of bricks to be a^2/m, a^2/n, a^2/p, a^2/q. If x, y, z, u be the number of bricks of each variety respectively that are employed in a layer, the problem amounts algebraically to the solution of the indeterminate equations

$$\left.\begin{aligned} \frac{x}{m}+\frac{y}{n}+\frac{z}{p}+\frac{u}{q} &= 7\tfrac{1}{2}, \\ x+y+z+u &= 200. \end{aligned}\right\} \qquad \ldots \quad \text{(B)}$$

Baudhāyana gives two integral solutions of these equations :[1]

(*i*) $m=16, \ n=25, \ p=36, \ q=100$

(*i*·1) $x=24, \ y=120, \ z=36, \ u=20$

or

(*i*·2) $x=12, \ y=125, \ z=63, \ u=0.$

Since in this method the values of m, n, p, q are perfect squares, the shape of all the bricks are square. Baudhāyana has described a second method of construction of the *Agni* in which he employs certain rectangular bricks too. These bricks, it may be noted, are easily divisible into square shapes. But as that will increase the number of bricks employed in the construction of the *Agni* he has refrained from doing so. All those things are, however, immaterial for us who look upon his problem from the point of view of the solution of algebraic equations. Baudhāyana's new solutions of the equations (B) are [2]

(*ii*) $m=25, \quad n=50, \quad p=50/3, \quad q=100$

(*ii*·1) $x=160. \quad y=30, \quad z=8, \quad u=2$

or

(*ii*·2) $x=165, \quad y=25, \quad z=6, \quad u=4$

Āpastamba uses square bricks of five different varieties for the construction of the same *Agni*. So his problem will be represented algebraically by the equations

$$\left.\begin{aligned} \frac{x}{m}+\frac{y}{n}+\frac{z}{p}+\frac{u}{q}+\frac{v}{r} &= 7\tfrac{1}{2}, \\ x+y+z+u+v &= 200\,; \end{aligned}\right\} \qquad \ldots \quad \text{(C)}$$

[1] *BSl*, iii. 24 ff.

[2] *BSl*, iii. 41 ff.

where m, n, p, q, r are perfect squares. Āpastamba's statement of his method of solution, partly explained in detail and partly hinted, is not free from ambiguity.[1] Consequently it has been interpreted differently by his different commentators leading consequently to several solutions. According to Karavindasvāmī, the two solutions of (C) will be

$$(i)\ \begin{cases} m=16, & n=25, \quad p=64, \quad q=100, \quad r=144, \\ x=67, & y=58, \quad z=48, \quad u=18, \quad v=9; \end{cases}$$

$$(ii)\ \begin{cases} m=16, & n=25, \quad p=36, \quad q=64, \quad r=100, \\ x=12, & y=157, \quad z=9, \quad u=0, \quad v=22. \end{cases}$$

To these Kapardisvāmī adds the solution

$$(iii)\ \begin{cases} m=16, & n=25, \quad p=36, \quad q=64, \quad r=100, \\ x=10, & y=159, \quad z=9, \quad u=8, \quad v=14. \end{cases}$$

Sundararāja's interpretation leads to as many as four new solutions of (C)

$$(iv), (v)\ \begin{cases} m=16, \quad n=25, \quad p=36, \quad q=64, \quad r=100, \\ x=70,\ 12; \ y=45,\ 157; \ z=9, \ u=56,\ 0; \ v=20,\ 22; \end{cases}$$

$$(vi), (vii)\ \begin{cases} m=16, \quad n=25, \quad p=64, \quad q=100, \quad r=144, \\ x=74,\ 77; \ y=45,\ 42; \ z=52,\ 40; \ u=20,\ 32; \ v=9 \end{cases}$$

He has also added a few more solutions of his own. All these show very clearly that the Hindus fully recognised the indeterminate character of the above problem of the construction of the Fire-altar of the shape of the falcon.

[1] *ĀpŚl*, xi. 1 ff. The text describing one solution is positively faulty as has also been noticed by all the commentators. Bürk failed to detect this. He thinks erroneously that the preliminary arrangement for the second construction is comprised of 194 bricks, whereas it actually consists of 198 bricks (*vide* *ZDMG*, LVI, p. 366).

Several other indeterminate problems of the above type present themselves in connexion with the construction of the Fire-altars of other shapes. We need not dilate upon them here. It should, however, be noted that the actual difficulties of construction are much more than what will appear from the mere algebraic considerations. For the bricks will have to be arranged in such a configuration as to have the prescribed shape of the altar.

CHAPTER XV

Elementary Treatment of Surds

In the *Śulba*, we find elementary treatment of surds, particularly their addition, multiplication and rationalization. For the face, base and altitude of the *Sautrā-maṇikī-vedi* which is of the shape of an isosceles trapezium, are stated, it has been noted before,[1] to be respectively $24/\sqrt{3}$, $30/\sqrt{3}$ and $36/\sqrt{3}$ prakramas, or to be $8\sqrt{3}$, $10\sqrt{3}$ and $12\sqrt{3}$ prakramas. Hence it is clear that the ancient Hindus knew that

$$\frac{24}{\sqrt{3}} = 8\sqrt{3}, \quad \frac{30}{\sqrt{3}} = 10\sqrt{3}, \quad \frac{36}{\sqrt{3}} = 12\sqrt{3}.$$

It is also stated in general that if the side of a square be a, then the side of the square equal to the third part of it will be $\frac{1}{3}(a\sqrt{3})$ or $\sqrt{3}(\frac{a}{3})$.[2] That is,

$$\frac{a}{\sqrt{3}} = \frac{a\sqrt{3}}{3}.$$

Thus it appears that the rationalization of simple surds was known at that time.

The area of the above trapezium is stated to be 324 square prakramas. It must have been calculated with the help of the rule given in the *Śulba* for that purpose. So that

$$\frac{36}{\sqrt{3}} \times \frac{1}{2}\left(\frac{24}{\sqrt{3}} + \frac{30}{\sqrt{3}}\right) = \frac{36}{3} \times \frac{54}{2} = 324.$$

$$12\sqrt{3} \times \tfrac{1}{2}(8\sqrt{3} + 10\sqrt{3}) = 12 \times 3 \times 9 = 324$$

1 *Supra*, p. 153.

2 *ĀpŚl*, ii- 3; *BŚl*, i. 47; *KŚl* ii- 15-6; see also pp. 74 f. *supra*.

Again the dimensions of another isosceles trapezium (*Āśvamedhikī-vedi*) are stated thus: face $=24\sqrt{2}$, base $=33\sqrt{2}$, altitude $=36\sqrt{2}$ prakramas and area $=1944$ square prakramas. That is

$$36\sqrt{2} \times \tfrac{1}{2}(24\sqrt{2} + 30\sqrt{2}) = 1944.$$

In the *Śulba*, a surd is technically called *karaṇī*. Thus *dvi-karaṇī*-means $\sqrt{2}$, *tri-karaṇī* $=\sqrt{3}$, *tṛtīya-karaṇī* $=\sqrt{1/3}$, *saptama-karaṇī* $=\sqrt{1/7}$, *aṣṭādaśa-karaṇī* $=\sqrt{18}$, etc.[1] The term was also used in the more general sense of a root. For we have at least one instance of its application in that sense, *e.g.*, *catuṣkaraṇī* $=\sqrt{4}$, which is not a surd.[2] The Sanskrit word *karaṇī* means "producer," "that which makes." From that it came to denote the sides of a rectilinear geometrical figure of any shape,[3] and then more particularly, the side of a square.

Approximate Value of $\sqrt{2}$.

Baudhāyana and Āpastamba say :

"Increase the measure (of which the *dvi-karaṇī* is to be found) by its third part, and again by the fourth part (of this third part) less by the thirty-fourth part of itself (*i.e.*, of this fourth part). (The value thus obtained is called) the *saviśeṣa*."[4]

Kātyāyana defines the rule in nearly identical words.[5] Thus if d be the *dvi-karaṇī* of a, that is, if d be the side

1 The references are respectively to *ĀpŚl*, i. 5, ii. 2, ii. 3 ix. 5, xix. 1.

2 *ĀpŚl*, ii. 6.

3 See *ĀpŚl*, ix. 6; xii. 5, 6, 9; xiii. 1, etc.

4 "प्रमाणं तृतीयेन वर्धयेत्तच्च चतुर्थेनात्मचतुस्त्रिंशोनेन। सविशेषः।"—*BŚl*, i. 61-2; *ĀpŚl*, i. 6.

5 "करणीं तृतीयेन वर्धयेत्तच्च स्वचतुर्थेनात्मचतुस्त्रिंशोनेन सविशेष इति विशेषः" —*KŚl*, ii. 13.

of a square whose area is double that of the square on a, then, according to the rule,

$$d = a + \frac{a}{3} + \frac{a}{3.4} - \frac{a}{3.4.34}.$$

Now, it has been stated before, the diagonal of a square is its *dvi-karaṇī*. So this rule gives the relation between the diagonal and side of a square. Indeed the above rule is particularly meant to define that relation. Thus we get

$$\sqrt{2} = 1 + \frac{1}{3} + \frac{1}{3.4} - \frac{1}{3.4.34}.$$

In terms of decimal fractions, this works out $\sqrt{2} = 1{\cdot}4142156....$ According to modern calculation $\sqrt{2} = 1{\cdot}414213....$ Thus it is clear that the ancient Hindus attained, a very remarkable degree of accuracy in calculating an approximate value of $\sqrt{2}$.

Hypotheses about its Origin.

One will be naturally interested to know how the value of $\sqrt{2}$ was determined in that early time to such a high degree of approximation. Unfortunately the Hindus have not left any trace of the method adopted by them for the purpose. So it is very difficult to guess it. Thibaut has, however, propounded an ingenious hypothesis about it. He says: [1]

"The question arises: how did Baudhāyana or Āpastamba or whoever may have the merit of the first investigation, find this value? Certainly they were not able to extract the square root of 2 to six places of decimals; if they have been able to do so, they would have arrived at still greater degree of accuracy. I suppose that they arrived at their result by the following method which account for the exact degree of accuracy they reached.

[1] *Śulbasūtras*, pp. 13 ff.

"Endeavouring to discover a square the side and diagonal of which might be expressed in integral numbers they began by assuming two as the measure of a square's side. Squaring two and doubling the result they got the square of the diagonal, in this case = eight. They then tried to arrange eight, let us say again, eight pebbles, in a square; as we should say they tried to extract the square root of eight. Being unsuccessful in this attempt, they tried the next number, taking three for the side of a square; but eighteen yielded a square root no more than eight had done. They proceeded in consequence to four, five, etc. Undoubtedly they arrived soon at the conclusion that they would never find exactly what they wanted, and had to be contented with an approximation. The object was now to single out a case in which the number expressing the square of the diagonal approached as closely as possible to a real square number. I subjoin a list, in which the numbers in the first column express the side of the squares which they subsequently tried, those in the second column the square of the diagonal, those in the third the nearest square number.

1	2	1	11	242	256
2	8	9	12	288	289
3	18	16	13	338	324
4	32	36	14	392	400
5	50	49	15	450	441
6	72	64	16	512	529
7	98	100	17	578	576
8	128	121	18	648	625
9	162	169	19	722	729
10	200	196	20	800	784.

"How far the Sūtrakāras went in their experiments we are of course unable to say; the list up to twenty suffices for our purposes. Three cases occur in which the number expressing the square of the diagonal of a square differs only by one from a square number; 8—9; 50—49; 288—289; the last case being most favourable, as it involves the largest numbers. The diagonal of a square the side of which was equal to twelve, was very little shorter than seventeen ($\sqrt{289} = 17$). Would it then not be possible to reduce 17 in such a way as to render the square of the reduced number equal or almost equal to 288?

"Suppose they drew a square the side of which was 17 padas long, and divided it into $17 \times 17 = 289$ small squares. If the side of the square could now be shortened by so much, that its area would contain not 289 but only 288 such small squares, then the measure of the side would be exact measure of the diagonal of the square, the side of which is equal to 12 ($12^2 + 12^2 = 288$). When the side of the square is shortened a little, the consequence is that two sides of the square a stripe is cut off; therefore a piece of that length had to be cut off from the side that the area of the two stripes would be equal to one of the 289 small squares. Now, as square is composed of 17×17 squares, one of the two stripes cuts off a part of 17 small squares and the other likewise of 17, both together of 34 and since these 34 cut off pieces are to be equal to one of the squares, the length of the piece to be cut off the side is fixed thereby; it must be the thirty-fourth part of the side of one of the 289 small squares.

"The thirty-fourth part of thirty-four small squares being cut off, one whole small square would be cut off and the area of the large square reduced exactly to 288 small squares; if it were not for one unavoidable circumstance.

The two stripes which are cut off from two sides of the square, let us say the east side and the south side, intersect or overlap each other in the south-east corner and the consequence is, that from the small square in that corner not $\frac{2}{34}$ are cut off, but only $\frac{2}{34} - \frac{1}{34 \times 34}$. Thence the error in the determination of the value of the saviśeṣa. When the side of a square was reduced from 17 to $16\frac{33}{34}$ the area of the square of that reduced side was not 288, but $288 + \frac{1}{34 \times 34}$. Or putting it in a different way: taking 12 for the side of a square, dividing each of the 12 parts into 34 parts (altogether 408) and dividing the square into the corresponding small squares, we get $408 \times 408 = 166464$. This doubled is 332928. Then taking the saviśesha-value of $16\frac{33}{34}$ for the diagonal and dividing the square of the diagonal into the small squares just described, we get $577 \times 577 = 332929$ such small squares. The difference is slight enough.

"The relation of $16\frac{33}{34}$ to 12 was finally generalised into the rule: increase a measure by its third, this third by its own fourth less the thirty-fourth part of this fourth $\left(16\frac{33}{34} = 12 + \frac{12}{3} + \frac{12}{3 \times 4} - \frac{12}{3 \times 4 \times 34}\right)$. The example of the saviśeṣa given by commentators is indeed $16\frac{33}{34} : 12$; the case recommended itself by being the first in which the third part of a number and the fourth part of the third part were both whole numbers."

But a more simple and very plausible hypothesis will be that the expression for $\sqrt{2}$ was obtained in the following way:[1] Take two squares whose sides are of unit

[1] For a slightly different but less elegant procedure see Müller, p. 178. *loc. cit.*

length. Divide the second square into three equal strips

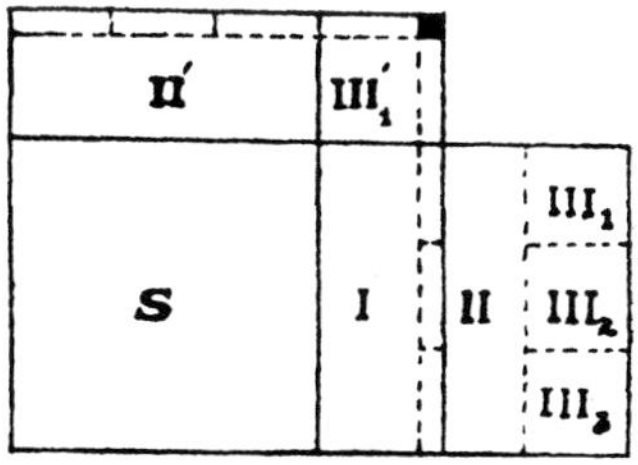

Fig. 75

I, II and III. Sub-divide the last strip into three small squares III_1, III_2, III_3 of sides $\frac{1}{3}$ each. Then on placing II and III_1 about the first square S in the positions II′ and III'_1, a new square will be formed. Now divide each of the portions III_2 and III_3 into four equal strips. Placing four and four of them about the square just formed, on its east and south sides, say, and introducing a small square at the south-east corner, a larger square will be formed, each side of which will be obviously equal to

$$1 + \frac{1}{3} + \frac{1}{3.4}.$$

Now this square is clearly larger than the two original squares by an amount $\left(\frac{1}{3.4}\right)^2$, the area of the small square introduced at the corner. So to get equivalence cut off from the either sides of the former square two thin strips. If x be the breadth of each thin strip, we must have

$$2x\left(1 + \frac{1}{3} + \frac{1}{3.4}\right) - x^2 = \left(\frac{1}{3.4}\right)^2.$$

Whence, neglecting x^2 as too small, we get

$$x = \left(\frac{1}{3.4}\right)^2 \cdot \frac{3.4}{34} = \frac{1}{3.4.34}.$$

Thus we have finally

$$\sqrt{2} = 1 + \frac{1}{3} + \frac{1}{3.4} - \frac{1}{3.4.34}.$$

nearly.

Approximate Value of $\sqrt{3}$.

By the process indicated above we can easily get an approximate value of $\sqrt{3}$. In this case two of the unit squares are divided into six equal strips I, II,...VI. The last two strips are subdivided into six smaller squares of sides 1/3 each. Then arranging the slices III, IV, V_1, V_2, VI_1, and VI_2, about the first square in the positions III′, IV′, V'_1, V'_2, VI'_1, VI'_2, a new square can be formed. Now divide each of the portions left over, *viz.*, V_3 and VI_3, into

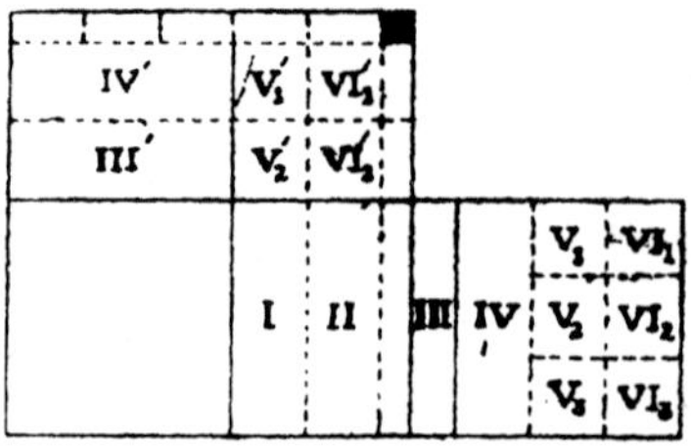

Fig. 76

five equal parts and place them about the square just formed. Then introducing a small square at the corner another complete square of sides equal to

$$1 + \frac{2}{3} + \frac{1}{3.5}$$

each will be formed. But this is clearly too large by the amount $(\frac{1}{35})^2$. So for closer approximation, let the side of the new square be diminished by an amount y, such that

$$2y\left(1 + \frac{2}{3} + \frac{1}{3.5}\right) - y^2 = \left(\frac{1}{3.5}\right).$$

Hence $y = \frac{1}{3.5.52}$, nearly.

Thus we get [1]

$$\sqrt{3} = 1 + \frac{2}{3} + \frac{1}{3.5} - \frac{1}{3.5.52}$$

Irrationality of $\sqrt{2}$.

Did the ancient Hindus recognise the irrationality of $\sqrt{2}$? This question does not seem to have troubled Thibaut in any way. For we do not find him to raise it explicitly or to attempt to answer it directly. But it is clear from his writings that he believed in the Hindu knowledge of irrationality of $\sqrt{2}$. Indeed his theory about the discovery of the Hindu approximation to the value of $\sqrt{2}$, which has been quoted *in extenso* before, is fundamentally based on the knowledge of the incommensurability of the diagonal of a square with its sides. Thibaut supposes that the ancient Hindus endeavoured 'to discover a square the side and diagonal of which might be expressed in integral numbers' and then observes, 'undoubtedly they arrived soon at the conclusion that they would never find exactly what they wanted and had to be contented with an approximation." Von Schroeder [2] and Bürk [3] are more explicitly emphatic. They claim for the ancient Hindus the credit for the first discovery of irrationals. And they have been followed in this respect by Garbe, Hopkins and Macdonell.[4] But this hypothesis has

[1] For a different method of approximating to the value of $\sqrt{3}$, see Müller, *loc. cit.*, pp. 182 f.

[2] Von Schroeder, *Pythagoras und die Inder*, Leipzig, 1884, pp. 39-59; compare also *Indien Literatur und Kultur*, Leipzig, 1887.

[3] *ZDMG*, LV, p. 557.

[4] R. Garbe, *Philosophy of Ancient India*, pp. 39 ff.; E. W. Hopkins, *Religion of India*, pp. 559 f.; A. A. Macdonell, *History of Sanskrit Literature*, p. 422.

been criticised and opposed by some modern historians of mathematics.[1]

There are two terms which have undoubtedly a great bearing on the point under discussion. They are *viśeṣa* and *saviśeṣa*. Importance of these terms has not been properly realised by previous writers. Thibaut simply observes that *saviśeṣa* is a technical name for the increased measure.[2] Bürk remarks, " The total increase is *viśeṣa* because it is the ' difference ' between the *pramāṇa*, *i.e.*, the side of the given square and its *dvi-karaṇī*. Therefore this latter is *saviśeṣa*, ' with the difference.' "[3]

In the *Śulba*, the calculated value of the diagonal of a square is technically called the *saviśeṣa* of its side. Or symbolically,[4]

$$\textit{Saviśeṣa} \text{ of } a = a + \frac{a}{3} + \frac{a}{3.4} - \frac{a}{3.4.34}$$

Now what is the radical significance of the term *saviśeṣa*? Before answering this question, we shall point out that occasionally the term has been used to denote the complete diagonal in general; that is, in the sense *saviśeṣa* of $a = a\sqrt{2}$.[5] Again in one instance in the *Āpastamba*

[1] H. G. Zeuthen, " Theorem de Pythagore. Origine de la géométrié scientifique," *Comptes Rendus du IIme Congrés Intr. d. Philosophie*, Genéve, 1904 ; M. Cantor, " Über die älteste indische Mathematik," *Arch. d. Math. u. Phys.*, VIII (3), 1905, pp. 63-72 ; H. Vogt, " Haben die alten Inder den Pytha goreischen Lehrsatz und das Irrationale gekannt ? " *Bibl. Math.*, VII (3), 1906, pp. 6-23 ; T. L. Heath, *Euclid*, I, p. 362 f.

[2] Thibaut, *Śulbasūtras*, p. 13.

[3] Bürk, *ZDMG*, LVI, p. 330 ; compare also LV, p. 548 (" *saviśeṣa*, *i.e.*, of the approximation to the *dvi-karaṇī* ") and p. 557, fn. 1.

[4] *BŚl*, i. 61-2 ; *ĀpŚl*, i.6 ; *KŚl*, ii. 13.

[5] *BŚl*, iii. 67, 149, 150 ; *ĀpŚl*, xix. 2, 3, 4, 7.

Śulba,[1] we find the use

$$\textit{Viśeṣa} \text{ of } a = \frac{a}{3} + \frac{a}{3.4} - \frac{a}{3.4.34}$$

But on several occasions in this work and also in other *Śulbas*,[2] particularly in a compound word, the term *viśeṣa* has been employed in the sense of the hypotenuse of a right-angled triangle. And it has again been considered there as equivalent to *saviśeṣa*.[3] The concluding portion of Kātyāyana's statement of the rule for *saviśeṣa* runs thus:

1 "पृष्ठान्तयोर्मध्ये च शङ्कुन्निहत्य अर्द्धं तद्विशेषमभ्यस्य लक्षणं कृत्वा अर्द्धमागमयेत् अन्त्यौ पार्श्वौ कृत्वा, मध्यमे सविशेषं प्रतिमुच्य ... " —*ĀpŚl*. ii. 1.

2 *BŚl*, iii. 164; *ĀpŚl*, xx. 5, 7, 8, 11, etc. In these cases occur the term *bāhya-viśeṣa* (meaning "having the *viśeṣa* outwards"); in *ĀpŚl*, xx. 6 we find *abhyantara-viśeṣa* ("having the *viśeṣa* inwards").

3 For example, one kind of bricks employed in the construction of the Fire-altar of the shape of falcon with bent wings and spreadout tail (*Vakrapakṣa-vyastapuccha-śyenacit*) was called *ṣoḍaśī*. Its dimensions are described thus:

"षोडशीं चतुर्भिः परिगृह्णीयात् अष्टमेन, त्रिभिरष्टमैश्चतुर्थेन चतुर्थसविशेषेणेति"—*ĀpŚl*, xix. 2.

"Construct the *ṣoḍaśī* with four (sides), namely with one-eighth, three-eighth, one-fourth (puruṣa) and the *saviśeṣa* of the one-fourth (puruṣa)." The manner of laying out these bricks has been described thus:

"अवशिष्टं षोडशीभिः परिच्छादयेत् अन्त्या बाह्यविशेषा अन्तरशिरसः"—*ĀpŚl*, xx. 5.

"Cover the remaining portion (of the Fire-altar) with *ṣoḍaśīs*, (such that) those lying at the extremity (of the Fire-altar) will have their *viśeṣa* outwards, but inwards at the head."

"अपरस्मिन् प्रस्तारे पुरस्ताच्छिरसि द्वे षोडश्यौ बाह्यविशेषे उपदध्यात् ते अपरेण द्वे विषये अभ्यन्तरविशेषे"—*ĀpŚl*, xx. 6.

"In the second layer, at the head towards the east lay two *ṣoḍaśīs* having their *viśeṣa* outwards and on the west lay them with their *viśeṣa* inwards and lying in both places (*viṣaya*, *i.e.*, partly in the head and partly in body of the Fire-altar)."

saviśeṣa iti viśeṣaḥ. Here the word *viśeṣa* must be explained differently. The use of the same terms thus in varied significations, has made the interpretation of their origin really difficult. Thibaut's explanation is obviously erroneous; Bürk's is not full.

Let us now see how the origin and significance of the term *saviśeṣa* has been interpreted by the early commentators. Dvārakanātha Yajvā is of no help to us in this matter. He passes off with the simple remark that "*saviśeṣa* is a technical term for it." Kapardisvāmī observes:

"*Saviśeṣa* is a technical name for the sum thus obtained.[1] As it is accompanied with a special quantity in excess (*viśeṣa*), so it is a term whose meaning is intelligible by itself. (Add) to twelve (aṅgulis) four (aṅgulis), to four one; divide that one into thirty-four parts and leave out a part. Thus (the resulting sum) will be seventeen aṅgulis less one *tila*. The square of the tilas in twelve aṅgulis is 166464 square tilas; the square of the tilas in seventeen aṅgulis minus one tila is 332929 square tilas. In this (latter) a square of one tila is in excess (of twice the other), so the *saviśeṣa* is the technical name (for it). If it is accompanied with some quantity in excess (*viśeṣa*), then what is the necessity for it? In its own domain, it has no fault. For if the diagonal be measured (directly) with a bamboo-stick, the excess will be to the extent of ten square tilas. Thus in any case there will be an excess even by a fraction of the smallest part of the minute *nīvāra* grain falling from the mouth of a parrot. So (the formula) is without fault. It will be of practical use, taking this into consideration, the learned author has made that (technical) term."

[1] The reference is to $a + \frac{a}{3} + \frac{a}{3.4} - \frac{a}{3.4.34}$

Karavindasvāmī's observations are more elaborate.

"*Saviśeṣa* is its technical name. The measure of the side (of a square) after having been operated upon by the rule 'increase the measure by its one-third, etc.' is called the *saviśeṣa*. For instance, increase a measure of twelve aṅgulis by four aṅgulis ; increase that four aṅgulis by one aṅguli minus one tila. The experts (in measure) say that thirty-four tilas placed breadthways make one aṅguli. It will be stated (later on in the text) 'add to half (the length of the east-west line) its *viśeṣa*,'[1] 'enclose with two half-bricks having their *viśeṣa* outwards,'[2] etc. What (are the correct meanings) of 'fasten *saviśeṣa* to the middle (pole),'[3] 'with one-fourth puruṣa and with the *saviśeṣa* of one-fourth puruṣa,'[4] etc. There it should be understood that 'being with the *viśeṣa*,' that is 'the measure with the *viśeṣa*' is the *saviśeṣa*. What is the need of this big term? So that its true significance may be intelligible by itself. What is that? The root *śiṣ* when prefixed by *vi* denotes in all cases 'a correction in excess.' As the rule mentions of an excess qualified by the prefix *vi* (it should be understood that the accurate value of) the diagonal differs by something from, exceeds over the exceeding part of the above value of the *dvi-karaṇī* over the measure of the side: and that is the *viśeṣa*. Or else the *viśeṣa* is that little area by which (the square of the diagonal as calculated by the rule) differs from or exceeds over (twice the area of the given square) at the time of measurement. For instance, since it has been stated that thirty-four tilas make one aṅguli, the square of the number of tilas in twelve aṅgulis is 166464 square areas of tilas ; the square of the tilas in seventeen aṅgulis minus one tila is 332929 square areas of tilas. In the latter a square of one tila is present in excess over twice the area of the (given)

1 *ĀpŚl*, ii. 1.
2 *Ibid*, xx. 7.
3 *Ibid*, ii. 1.
4 *Ibid*, xix. 2.

square figure ; hence its technical name is *viśeṣa*. If the diagonal (of the square) could be (exactly) measured with the bamboo-rod and the Fire-altar had been measured by means of that measure, the area included in the body (of the Fire-altar) would not have exceeded (twice) this (the given square) even by the smallest part of the minute *nīvāra* grain falling from the mouth of a parrot. If the measure falls short (the area described) in that case will also be smaller. So for doubling (a square), the *saviśeṣa* is employed as a practical means."

Thus it is evident that according to the interpretation of the commentators Kapardisvāmī and Karavindasvāmī, the term *saviśeṣa* for the diagonal of a square implies intrinsically a knowledge of the following : (1) The value of the diagonal as calculated by the rule stated for the purpose is only an approximate one; (2) that value of the diagonal has a small quantity in excess over the true value of the diagonal; or in other words, the square of that value exceeds by some quantity twice the area of the given square; (3) that excess cannot be completely eliminated in calculating the value of the diagonal arithmetically. If the validity of this interpretation be accepted, then there will remain nothing to doubt that the ancient Hindus were aware of the incommensurability of the diagonal of a square with its sides.

Let us therefore test as far as possible how far the commentators can be relied upon in the matter of that interpretation. For it might be argued by modern critics that being acquainted with the real state of things from the mathematical knowledge of their time, the commentators naturally gave an interpretation to the texts which was more creditable to their authors.[1] Looking into the

[1] The commentators have been similarly accused in other connexions by Thibaut (*Śulbasūtras*, pp. 46 f.) and Bürk (comments on *ĀpŚl*, i. 4).

ancient literatures of India, we find in the early canonical works of the Jainas copious instances of the employment of the term *viśeṣa* in the same connection as we find it in the *Sulba*. Thus in the *Sūryaprajñapti* (c. 500 B. C.)[1] the circumference of a circle whose diameter is 99640 yojanas is stated as 315089 yojanas and a little over (*kiñcidviśeṣādhika*); that of a circle of diameter 100660 yojanas is stated to be 318315 and a little less (*kiñcid-viśeṣoṇa*). In the *Jambudvīpaprajñapti* (c. 300 B.C.),[2] the circumference of the Jambudvīpa which is of the shape of a circle of 100000 yojanas in diameter, is mentioned as 316227 yojanas 3 gavyutis 128 dhanus $13\frac{1}{2}$ aṅgulis and a little over (*kiñcid-viśeṣādhika*). In all these cases,[3] it will be seen that the value actually recorded is only an approximate one and the *viśeṣa* refers to a small quantity which has not been recorded—in fact it cannot be accurately determined—but which we shall have to add to or subtract from the recorded value in order to get the accurate value of the quantity sought. Those who are acquainted with the technique of Sanskrit and Sanskritic languages will at once recognise that the expression *kiñcid-viśeṣādhika* has the identical significance as the term *saviśeṣa*. Indeed we really find the use of the word *saviśeṣa* in exactly the same sense in a Jaina work of later times. Nemicandra (c. 975 A. D.) writes: *tattiuṇaṁ parirayeṇa saviseṣaṃ* or "thrice that with a little over (*saviśeṣa*) is the circumference."[4] Now the early canonical works of

[1] Sūtra 20. The formula employed for the calculation is

$$\text{circumference} = \sqrt{10 \times (\text{diameter})^2}.$$

[2] Sūtra 3.

[3] Several instances of this kind will be found in the author's article on "The Jaina School of Mathematics" in the *Bull. Cal. Math. Soc.*, Vol. XXI, 1929, pp. 115-145; see particularly pp. 131-3.

[4] *Trilokasāra* of Nemicandra, with the commentary of Mādhavacandra, edited by Manoharlal Sastri, Bombay, 1918, Gāthā 95.

the Jainas belong to a period not much separated from that of the *Śulba*. Some works of either classes probably belong to the same age. So we can accept without any hesitation that the term *saviśeṣa* was employed originally in the *Śulba* with the same significance as that with which it is found to have been employed in the early canonical works of the Jainas. Thus it is found conclusively that Kapardisvāmī and Karavindasvāmī are thoroughly reliable as regards their interpretation of the original significance of the term *saviśeṣa*. So it is proved that the irrationality of $\sqrt{2}$ was known to the ancient Hindus.

Other Approximate Values of $\sqrt{2}$.

From the *Mānava Śulba* we obtain certain other noteworthy approximations to the value of $\sqrt{2}$. By way of some calculations in that work are employed the relations:

$$(1) \qquad 40^2 + 40^2 = 56^2$$
$$(2) \qquad 4^2 + 4^2 = (5\tfrac{2}{3})^2$$

From these we easily derive the values,

$$(1{\cdot}1) \qquad \sqrt{2} = \tfrac{7}{5} = 1{\cdot}4$$
$$(2{\cdot}1) \qquad \sqrt{2} = \tfrac{17}{12} = 1{\cdot}4166\ldots$$

It may be noted that $\frac{7}{5}$ is the third convergent of $\sqrt{2}$ expressed as a continued fraction [1] and the value $\frac{17}{12}$ is its fourth convergent. What is still more noteworthy is the fact that the former value is not derivable from the series for $\sqrt{2}$ stated before. This latter is, however, the eighth convergent of the continued fraction for $\sqrt{2}$:

$$\frac{577}{408} = 1 + \frac{1}{3} + \frac{1}{3.4} - \frac{1}{3.4.34}.$$

[1] $\sqrt{2} = 1 + \frac{1}{2+}\ \frac{1}{2+} \ldots$

These facts will lead one strongly to suspect if the rudiments of the theory of continued fractions were known to the early Hindus. At any rate, these are very remarkable cases of coincidence.

There are certain other rough values of $\sqrt{2}$ to which we shall refer shortly.

Approximation to the Value of $\sqrt{5}$.

There seems to have been a serious attempt, though without much success, to find an approximation to the value of the surd $\sqrt{5}$. The occasion was to define clearly the relative positions of the three principal and oldest known fire-altars, *viz.*, the *Gārhapatya*, *Āhavanīya* and *Dakṣiṇa*. Baudhāyana's rules to determine their positions are these:

"With the third part of the length (*i.e.*, the distance between the *Gārhapatya* and *Āhavanīya*) describe three squares closely following one another (from the west towards the east); the place of the *Gārhapatya* is at the north-western corner of the western square; that of the *Dakṣiṇāgni* is at its south-eastern corner; and the place of the *Āhavanīya* is at the north-eastern corner of the eastern square." [1]

'Or else divide the distance between the *Gārhapatya* and *Āhavanīya* into five or six (equal) parts; add (to it) a sixth or seventh part; then divide (a cord as long as) the whole increased length into three parts and make a mark at the end of two parts from the eastern end (of the cord). Having fastened the two ends of the cord (to the two) poles at the extremities of the distance between the *Gārhapatya* and *Āhavanīya*, stretch it toward the south, having taken it by the mark and fix a pole at the point reached. This is the place of the *Dakṣiṇāgni*." [2]

[1] *BŚl*, i. 67. [2] *BŚl*, i. 68.

"Or else increase the measure (between the *Gārhapatya* and *Āhavanīya*) by its fifth part; divide (a cord as long as) the whole into five parts and make a mark at the end of two parts from the western extremity (of the cord). Having fastened the two ties at the ends of the east-west line, stretch the cord towards the south having taken it by the mark and fix a pole at the point reached. This is the place of the *Dakṣiṇāgni*."[1]

The second is also given by Āpastamba.[2] A rule leading to the same result as the first one above, though defined differently, is stated by Kātyāyana[3] and Manu.[4] Kātyāyana has specified the relative positions of the three fire-altars also in a new way:

"Divide the distance between the *Gārhapatya* and *Āhavanīya* into six or seven parts; add a part; then divide (a cord) equal to the total increased length into three parts, etc."[5]

The rest of this rule is the same as the latter portion of the second rule above and hence need not be mentioned.

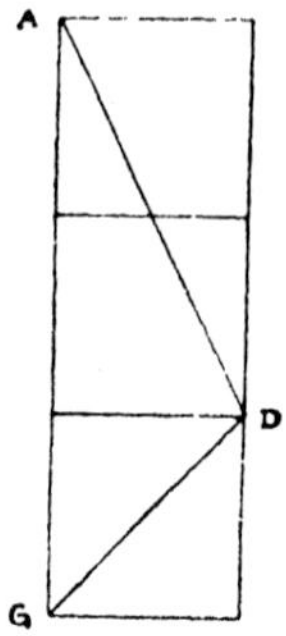

A = *Āhavanīya*
G = *Gārhapatya*
D = *Dakṣiṇāgni*

Fig. 77

1 *Ibid*, i. 69. 2 *ĀpŚl*, iv. 4. 3 *KŚl*, i. 29.
4 *MāŚl*, iii. 5 *KŚl*, i. 27.

Let b denote the distance between the *Gārhapatya* and *Āhavanīya*, that is, AG. Then from the different specifications given above we obtain the following values for AD and GD:

$$AD = \tfrac{b}{3}\sqrt{5},\ \tfrac{4}{5}b,\ \tfrac{7}{9}b,\ \tfrac{16}{21}b,\ \tfrac{18}{25}b$$

$$GD = \tfrac{b}{3}\sqrt{2},\ \tfrac{2}{5}b,\ \tfrac{7}{18}b,\ \tfrac{8}{21}b,\ \tfrac{12}{25}b$$

If it be assumed that the relative positions of the three fire-altars were meant to be the one and the same, in all cases though expressed differently, then we shall have the following approximations to the values of $\sqrt{5}$ and $\sqrt{2}$:

$$\sqrt{5} = 2\tfrac{2}{5},\ 2\tfrac{1}{3},\ 2\tfrac{2}{7},\ 2\tfrac{4}{25},$$

$$= 2{\cdot}4,\ 2{\cdot}333\ldots,\ 2{\cdot}285\ldots,\ 2{\cdot}16.$$

$$\sqrt{2} = 1\tfrac{1}{5},\ 1\tfrac{1}{6},\ 1\tfrac{1}{7},\ 1\tfrac{11}{25},$$

$$= 1{\cdot}2,\ 1{\cdot}166\ldots,\ 1{\cdot}142\ldots,\ 1{\cdot}44.$$

Since according to modern calculation $\sqrt{2} = 1{\cdot}414213\ldots$ and $\sqrt{5} = 2{\cdot}23607\ldots$, none of the above values can be said to be a fair approximation, perhaps except the values $\sqrt{5} = 2\tfrac{2}{7}$ and $\sqrt{2} = 1\tfrac{11}{25}$ which are correct up to the first place of decimals.

Evaluation of Other Surds.

In the *Mānava Śulba*, we find results leading incidentally to the evaluation of two other surds:

$$36^2 + 90^2 = 97^2$$

$$5^2 + 61^2 = (7\tfrac{5}{6})^2$$

Whence we easily obtain

$$\sqrt{29} = 5\tfrac{7}{18} = 5{\cdot}3888\ldots$$

$$\sqrt{61} = 7\tfrac{5}{6} = 7{\cdot}8333\ldots$$

Now correct up to three places of decimals $\sqrt{29} = 5 \cdot 385...$ and $\sqrt{61} = 7 \cdot 810...$ Hence the above approximations, specially the first one, may be said to be fair for ordinary purpose.

Approximate Formula.

It would be natural to ask if there is in the *Śulba* any specific rule for determining the approximate value of any surd. But the answer is not very reassuring. For we do not find an express statement of any formula for the purpose. At the same time we can unhesitatingly admit that they had the necessary equipments for the approximate evaluation of surds, at least of some. We have pointed out before that in the science of the *Śulba*, it is sometimes necessary to construct a square having a given area. And that is a geometrical method of finding the square-root of a given number. If the given area is represented by a non-square number, we get a method finding the square-root of a non-square number. One formula follows at once from the method given in the *Śulba*, for the transformation of a rectangle into a square, which has been described before. According to this method, a square (having its side equal to the breadth of the given rectangle) is first cut off from the given rectangle; the excess portion is divided into two halves which are then joined to the two sides of that square. Then by adding a small square in the corner, a large square is completed. The square equivalent to the given rectangle will be obtained, it is said, by subtracting the added small square from the completed large square. This subtraction was doubtless made by the Śulba-workers with the help of the theorem of the square of the diagonal. But it can also be made by cutting off two strips from the two sides, say, the east side and the south side of the completed large square.[1] Suppose A to be the area of the given rectangle and a be the side of the square subtracted from it. Then the side of the large completed square will be $a + \frac{A-a^2}{2a}$ The area of the small added square will be $\left(\frac{A-a^2}{2a}\right)^2$ Then from each side of the large sphere we shall have to cut off a thin strip of the same breadth. If x denote the breadth of the strip, we must have

$$\left(\frac{A-a^2}{2a}\right)^2 = 2x\left(a + \frac{A-a^2}{2a}\right) - x^2$$

or
$$x = \frac{\left(\frac{A-a^2}{2a}\right)^2}{2\left(a + \frac{A-a^2}{2a}\right)}$$

neglecting x^2 as being very small. Hence we finally arrive at the formula

$$\sqrt{A} = a + \frac{A-a^2}{2a} - \frac{\left(\frac{A-a^2}{2a}\right)^2}{2\left(a + \frac{A-a^2}{2a}\right)}$$

This formula requires a correction, it will be easily recognised, inasmuch as a portion equivalent to x^2 has been subtracted too much.

[1] The process is in fact the reverse of that taught in the *Sulba* for the increment of a given square into another square.

The rule of the Bakhshālī Manuscript for determining the approximate root of a non-square number must have been obtained exactly in this way. It says [1]

"In case of a non-square (number), subtract the nearest square number; divide the remainder by twice (the root of that number). Half the square of that (that is, the fraction just obtained) is divided by the sum of the root and the fraction and subtract; (this will be the approximate value of the root) less the square (of the last term)."

If $A = a^2 + r$, we write the above formula as

$$\sqrt{A} = \sqrt{a^2+r} = a + \frac{r}{2a} - \frac{(r/2a)^2}{2(a + r/2a)}$$

Now this formula will not be available for finding the approximate values of those surds in which r is not small compared with a^2. So Rodet [2] holds that a different process of approximation to the value of a surd was also known to the authors of the *Śulba*. It will lead to the formula, says he,

$$\sqrt{a^2+r} = a + \frac{r}{2a+1} + \frac{\frac{r}{2a+1}\left(1 - \frac{r}{2a+1}\right)}{2\left(a + \frac{r}{2a+1}\right)} + \epsilon$$

1 Bibhutibhusan Datta, "The Bakhshālī Mathematics," *Bull. Cal. Math. Soc.*, XXI, 1929, pp. 1-60.

2 L. Rodet, "Sur une méthode d'approximation des recines cares conne dans l'Inde antérieurment a la conquite d'Alexandre, *Bull. Soc. Math. d. France*, VII, 1879, pp. 98-102; "Sur les méthodes d'approximation chez las anciens."—*Ibid*, pp. 159-167.

where

$$\epsilon = \left[r - \left\{ \frac{r}{2a+1} + \frac{\frac{r}{2a+1}\left(1 - \frac{r}{2a+1}\right)}{2\left(a + \frac{r}{2a+1}\right)} \right\} \right.$$

$$\left. \times \left\{ 2a + \frac{r}{2a+1} + \frac{\frac{r}{2a+1}\left(1 - \frac{r}{2a+1}\right)}{2\left(a + \frac{r}{2a+1}\right)} \right\} \right]$$

$$\div \; 2\left\{ a + \frac{r}{2a+1} + \frac{\frac{r}{2a+1}\left(1 - \frac{r}{2a+1}\right)}{2\left(a + \frac{r}{2a+1}\right)} \right\}$$

A nearly equivalent formula will be obtained by proceeding in the following way: From the given rectangle A, cut off the square (S) of side a each. Let the area of the remaining portion of the rectangle be r. From this cut off two strips I and II of the same breadth $r/(2a+1)$ and arrange them as in Fig. 78. Then the area of the strip of the rectangle still left over will be

$$= r - 2a \times \frac{r}{2a+1} = \frac{r}{2a+1}.$$

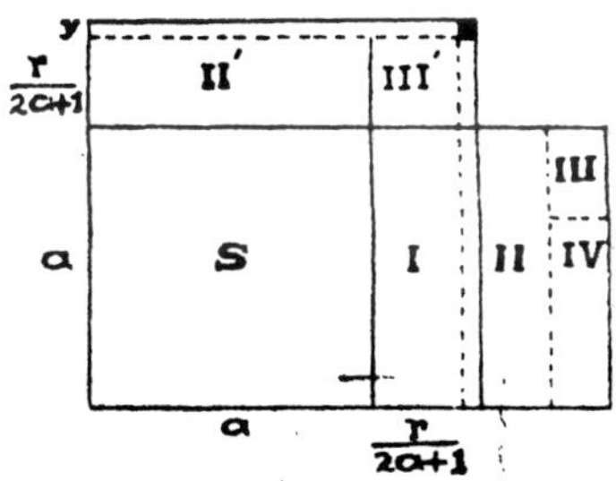

Fig. 78

Complete the larger square by adding a portion III which is deducted from $r/(2a+1)$. Then the area of IV will be

$$= \frac{r}{2a+1} - \left(\frac{r}{a+1}\right)^2 = \frac{r}{2a+1}\left(1 - \frac{r}{2a+1}\right).$$

Now this portion can be utilised in increasing the side of the square obtained before. If the increment of the side be y, then we must have

$$2y\left(a + \frac{r}{2a+1}\right) + y^2 = \frac{r}{2a+1}\left(1 - \frac{r}{2a+1}\right).$$

Therefore, approximately

$$y = \frac{\frac{r}{2a+1}\left(1 - \frac{r}{2a+1}\right)}{2\left(a + \frac{r}{2a+1}\right)}$$

Hence the side of the equivalent square is nearly

$$a + \frac{r}{2a+1} + \frac{\frac{r}{2a+1}\left(1 - \frac{r}{2a+1}\right)}{2\left(a + \frac{r}{2a+1}\right)}$$

This is a little too great. So decrease the square by cutting off a strip of breadth ϵ from either side; then

$$2\epsilon\left\{a + \frac{r}{2a+1} + \frac{\frac{r}{2a+1}\left(1 - \frac{r}{2a+1}\right)}{2\left(a + \frac{r}{2a+1}\right)}\right\} - \epsilon^2$$

$$= \left\{\frac{\frac{r}{2a+1}\left(1 - \frac{r}{2a+1}\right)}{2\left(a + \frac{r}{2a+1}\right)}\right\}^2$$

whence, we get

$$\epsilon = \left\{ \frac{\frac{r}{2a+1}\left(1-\frac{r}{2a+1}\right)}{2\left(a+\frac{r}{2a+1}\right)} \right\}^2$$

$$\div 2\left\{ a+\frac{r}{2a+1}+\frac{\frac{r}{2a+1}\left(1-\frac{r}{2a+1}\right)}{2\left(a+\frac{r}{2a+1}\right)} \right\}$$

Thus we have finally

$$\sqrt{A}=\sqrt{a^2+r}=a+\frac{r}{2a+1}+\frac{\frac{r}{2a+1}\left(1-\frac{r}{2a+1}\right)}{2\left(a+\frac{r}{2a+1}\right)}-\epsilon$$

nearly.

CHAPTER XVI

FRACTIONS AND OTHER MINOR MATTERS

Terminology.

In the *Sulba*, as in later Hindu mathematics, the fraction is called *aṃśa*, *bhāga*, meaning "part," "portion." Once in the *Āpastamba Sulba* and *Kātyāyana Sulba* each, it is called *kalā*.[1] This term is interesting inasmuch as it was used as early as the *Ṛgveda* [2] to denote particularly a sixteenth part. It is also employed symbolically for the number sixteen. The unit fraction is indicated by a compound of either of those terms with a cardinal number, *e.g.*, *pañcadaśa-bhāga* = 1/15 (*ĀpŚl*, x. 3; *KŚl*, v. 4); *tri-bhāga* = 1/3 (*KŚl*) or with an ordinal number, *e.g.*, *pañcama-bhāga* = 1/5 (*ĀpŚl*, ix. 7, x. 2; *KŚl*, v. 6). Oftentimes in the latter case, the word *bhāga* is omitted, so that we have only an ordinal to denote a fraction, *e.g.*, *pañcama* (or "fifth") = 1/5, *dvādaśa* ("twelfth") = 1/12, *trayodaśa* ("thirteenth") = 1/13, etc.[3]

In the *Mānava Sulba*,[4] we find a few very strange and unusual instances where *dvi-guṇa*, *tri-guṇa* and *catur-guṇa* are employed to denote, respectively 1/2, 1/3, and 1/4. But there we also meet with such usual use as *dvi guṇa* = 2-times, *pañca-guṇa* = 5-times.[5] The former are, indeed, highly ambiguous applications.

It is much noteworthy that the authors of the *Sulba* did not restrict themselves to the use of unit fractions only, as is known to have been the case with the early

[1] *ĀpŚl*, iii. 10; *KŚl*, iii. 11. Compare also *Chāndogya Upaniṣad*, vi. 7. 1.

[2] *ṚV*, viii. 47. 17.

[3] *BŚl*, i. 61, ii. 67; *ĀpŚl*, ix. 7, xii. 1, etc. It should be noted that *dvādaśa*, *ṣoḍaśa*, etc., are also used in the cardinal sense.

[4] *MāŚl*, v. 5.

[5] *Ibid*, ii. 5, 6.

Egyptian, Babylonian, and Chinese mathematicians.[1] In the *Śulba*, the unit fraction has not, indeed, any special significance attached to it. We find in them the frequent use of the general fraction. Their mode of expressing it is exactly the same as that of later Hindu writers. Thus 3/8 is called *tri-aṣṭama* (" three-eighths "), 2/7 *dvi-saptama* ("two-sevenths ").[2] Kātyāyana mentions $14\frac{3}{7}$ prakramas as *caturdaśa prakramān trīṃśca prakrama-saptabhāgān* (or " 14 prakramas and three of the seventh parts of a prakrama ").[3] 3/4 is sometimes called *catur-bhāgoṇa* (" less one-fourth "),[4] that is, $1-1/4$.

A peculiar mode of expressing certain fractions is sometimes found in the *Śulba*, *e.g.*, *ardha-navama*, which literally means " containing a half for its ninth," is used to denote " eight and a half "; *ardha-daśama* (" containing a half for its tenth) " $=9\frac{1}{2}$, and so on.[5] Such a term evidently carries with it the concrete concept of the operation of measuring.

A fraction of a fraction is indicated in the usual way thus: *jānoḥ pañcamasya caturviṃśena* = " by $\frac{1}{24}$ of $\frac{1}{5}$ of a *jānu*."[6] Further *caturtha-saviśeṣārdha* $= \frac{1}{2}\ (\frac{1}{4}\sqrt{2})$, *caturtha-saviśeṣa-saptama* $= \frac{1}{7}\ (\frac{1}{4}\sqrt{2})$.[7]

Operations with Fractions.

In the *Śulba*, there are instances showing fundamental arithmetical operations with elementary fractions. For example, it is stated in the *Baudhāyana Śulba*:[8]

[1] D. E. Smith, *History of Mathematics*, in two volumes, Boston, 1925; Vol. II, pp. 208 ff.

[2] *ĀpŚl*, xix. 2, 6. [3] *KŚl*, vi. 2. [4] *ĀpŚl*, xv. 5, xix. 1.

[5] *BŚl*, ii. 1-3; *ĀpŚl*. iii. 8; *MāŚl*, ii. 1-2.

[6] *BŚl*, ii. 13. [7] *ĀpŚl*, xix. 4, 7.

[8] *BŚl*, iii. 106. Compare also *BŚl*, iii. 238-9 and *ĀpŚl*, xviii. 3 which give

$$7\frac{1}{2} + \frac{1}{16} = 120.$$

"One hundred eighty-seven and a half square bricks of sides (equal to) one-fifth of a puruṣa make up the seven-fold *Agni* with the two aratnis and the prādeśa."

Here the area of each brick is 1/25 of a square puruṣa; so the number of such bricks required to cover an area of $7\frac{1}{2}$ square puruṣas will be

$$7\frac{1}{2} \div \frac{1}{25} = \frac{15}{2} \times 25 = 187\frac{1}{2}$$

Thus it is an instance of division of fractions. Or the same result may have been obtained in a slightly different way which is, indeed, a simplified method of division. Since the area of each brick is 1/25 of a square puruṣa, one square puruṣa will contain 25 such bricks. Therefore an area of $7\frac{1}{2}$ square puruṣas will contain

$$7\frac{1}{2} \times 25 = 187\frac{1}{2}$$

bricks. *If the area of each brick be* One-fifteenth of half of a square puruṣa," says Baudhāyana,[1] the number of bricks used will be 225. That is,

$$7\frac{1}{2} \div \frac{1}{15} \text{ of } \frac{1}{2} = \frac{15}{2} \div \frac{1}{30} = \frac{15}{2} \times 30 = 225.$$

In describing the *Droṇacit*, Baudhāyana writes:[2]

"Its body is a square; its side is three puruṣas less one-third. On the western side of the body is the handle. Its length east-to-west is half a puruṣa plus ten aṅgulis and its breadth north-to-south is one puruṣa less one-third. Thus is made the sevenfold *Agni* with the two aratnis and the prādeśa."

[1] *BŚl*. iii. 188-9. [2] *BŚl*, iii. 219-224.

That is,

$$7\frac{1}{2} \text{ square puruṣas} = \left\{\left(3 - \frac{1}{3}\right) \text{puruṣas}\right\}^2$$

$$+ \left(\frac{1}{2} \text{ puruṣa} + 10 \text{ aṅgulis}\right) \times \left\{\left(1 - \frac{1}{3}\right) \text{puruṣa}\right\};$$

the right-hand side, in square puruṣas

$$= \left(2\frac{2}{3}\right)^2 + \left(\frac{1}{2} + \frac{1}{12}\right)\left(1 - \frac{1}{3}\right),$$

$$= 7\frac{1}{9} + \frac{7}{12} \times \frac{2}{3},$$

$$= 7\frac{1}{2}.$$

The spatial dimensions of the constituent parts of the Fire-altar of the shape of the falcon with bent wings and spread-out tails have been described by Āpastamba as follow:

" Of the whole area making the seven-fold *Agni* with the two aratnis and the prādeśa, take off the prādeśa (from the tail), and the fourth part of the body together with eight quarter bricks. Of these latter, (use) three for the head, then divide the remainder between the two wings." [1]

Now, as is well-known, the body of the primitive Fire-altar of the shape of the falcon measures 4 square puruṣas, each wing $1 \times 1\frac{1}{5}$ square puruṣas and the tail $1 \times 1\frac{1}{10}$ square puruṣas. On taking out the prādeśa (= 1/10 puruṣa) from the tail, there will remain 1 square puruṣa. The body will be reduced by

$$4 \times \frac{1}{4} + 8 \times \frac{1}{16} \text{ square puruṣas};$$

[1] *ĀpŚl*, xv. 3.

there will then remain

$$4 - \left(4 \times \frac{1}{4} + 8 \times \frac{1}{16}\right) = 2\frac{1}{2} \text{ square puruṣas.}$$

Of the former with $3 \times \frac{1}{16}$ square puruṣas is formed the head and the remainder

$$\frac{4}{4} + \frac{8}{16} - \frac{3}{16} = 1\frac{5}{16} \text{ square puruṣas}$$

together with $1 \times \frac{1}{10}$ square puruṣas from the tail is divided equally between the two wings. Each wing will therefore measure

$$1 \times 1\frac{1}{5} + \frac{1}{2}\left(1\frac{5}{16} + \frac{1}{10}\right) \text{ square puruṣas.}$$

Hence the total area of the Fire-altar will be, in square puruṣas,

$$1 + 2\frac{1}{2} + \frac{3}{16} + 2\left\{1\frac{1}{5} + \frac{1}{2}\left(1\frac{5}{16} + \frac{1}{10}\right)\right\} = 7\frac{1}{2}$$

Of the squaring of a fraction, we take the following example from the *Āpastamba Śulba*:

"A cord $1\frac{1}{2}$ puruṣas long produces (a square of) $2\frac{1}{4}$ (square puruṣas); $2\frac{1}{2}$ puruṣas produce $6\frac{1}{4}$ (square puruṣas)."[1]

That is,

$$\left(1\frac{1}{2}\right)^2 = 2\frac{1}{4}, \quad \left(2\frac{1}{2}\right)^2 = 6\frac{1}{4}.$$

[1] *ĀpŚl*, iii. 8.

Baudhāyana states, on the contrary, that the side of a square measuring $7\frac{1}{9}$ square puruṣas is $2\frac{2}{3}$ puruṣas in length.[1] That is

$$\sqrt{7\tfrac{1}{9}} = 2\frac{2}{3}$$

Progressive Series.

In the manner of laying out bricks described in the *Śulba*, we find a few interesting instances of progressive series. Āpastamba writes:

" On (the occasion of) the first construction, (the altar-builder) should construct (the Fire-altar) knee-deep with 1000 bricks; on the second, navel-deep with 2000 bricks; on the third, mouth-deep with 3000 bricks. (The number of bricks employed in constructing the Fire-altar becomes thus) greater and greater on each successive occasion. He who constructs to attain the Heaven, (should thus construct with) great, high and unlimited (*mahāntam bṛhantam aparimitam*) (number of bricks); so it is known (from the ancient scriptures). "[2]

Thus we have the A. P.

1000, 2000, 3000,...

Reference to this progressive mode of successive construction of the Fire-altar is found as early as the *Taittirīya Saṃhitā* (c. 3000 B.C.).[3] It reappears in the *Śatapatha Brāhmaṇa* and Āpastamba expressly admits to have borrowed it from that work.[4] We find there another noteworthy instance which shows that it was very likely known then how to sum up a series in A. P.

[1] *BŚl*, iii. 220.

[2] *ĀpŚl*, x. 8; see also *ĀpŚr*, xvi. 13. 11-2; *BŚl*, ii. 26.

[3] *TS*, v. 6. 8. 2-3.

[4] *ĀpŚl*, xvi. 13. 12.

"But, indeed, that Fire-altar also is the Metres; for there are seven of these metres, increasing by four syllables; and the triplets of these make seven hundred and twenty syllables, and thirty-six in addition thereto." [1]

It has been stated elsewhere in the same work that the shortest metre is the *Gāyatrī* with 24 syllables. Thus we are given the first term (24), the common difference (4) and the number of term (7) of a series in A. P. Then its

$$\text{Sum} = \frac{7}{2}\ \{2 \times 24 + (7-1)4\}$$
$$= 252.$$

So the triplets of these metres will consist of 756 (= 252 × 3) syllables, which are equal to 720 + 36 as stated.

From the method indicated by Baudhāyana [2] for constructing larger and larger squares, starting with a smaller one, by adding successively gnomons to it is clear that the following series was known to him. [3]

$$1 + 3 + 5 + 7 + \cdots + (2n+1) = (n+1)^2$$

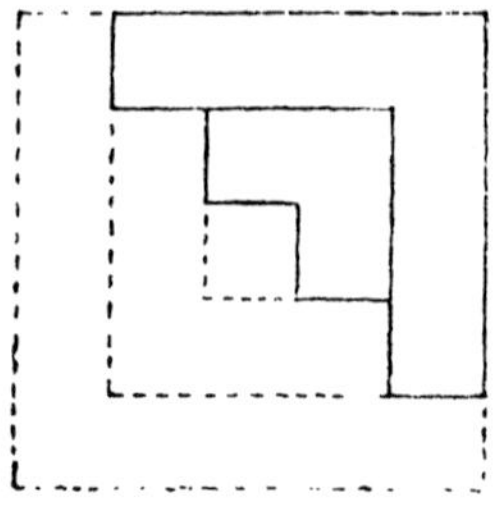

Fig 79

Factorisation.

The true significance of another passage is not quite clear to us unless it is some mystic expression of an

[1] *ŚBr*, x. 5. 4. 7. Eggeling's translation. [2] *Vide supra*, pp. 125 ff.
[3] Compare Müller, *loc. cit.*, pp. 200 f.

attempt to find all the possible factors of a number. The *Śatapatha Brāhmaṇa* says:

"Now in this Prajāpati, the year, there are 720 days and nights, his lights, (being) those bricks; 360 enclosing stones, and 360 bricks with (special) formulas. This Prajāpati, the year, has created all existing things, both what breathes and the breathless, both gods and men. Having created all existing things, he felt like one emptied out, and was afraid of death. He bethought himself, 'How can I get these beings back into my body? how can I put them back into my body? how can I be again the body of all these beings?' He divided his body into two; there were 360 bricks in the one, and as many in the other: he did not succeed. He made himself three bodies,—in each of them there were 3 × 80 of bricks: he did not succeed. He made himself four bodies of 180 bricks each: he did not succeed. He made himself five bodies,—in each of them there were 144 bricks: he did not succeed. He made himself six bodies of 120 bricks each: he did not succeed. He did not develop himself sevenfold.[1] He made himself 8 bodies of 90 bricks each: he did not succeed. He made himself 9 bodies of 80 bricks each: he did not succeed. He made himself 10 bodies of 72 bricks each: he did not succeed. He did not develop elevenfold. He made himself 12 bodies of 60 bricks each: he did not succeed. He did not develop either thirteenfold or fourteenfold. He made himself 15 bodies of 48 bricks each: he did not succeed. He made himself 16 bodies of 45 bricks each: he did not succeed. He did not develop seventeenfold. He made himself 18 bodies of 40 bricks each: he did not succeed. He did not develop nineteenfold.

[1] The text न सप्तधा व्यभवत् literally means "did not become divided into seven (parts)."

He made himself 20 bodies of 36 bricks each: he did not succeed. He did not develop either twenty-onefold, or twenty-twofold, or twenty-threefold. He made himself 24 bodies of 30 bricks each. There he stopped, at the fifteenth; and because he stopped at the fifteenth arrangement there are fifteen forms of the waxing, and fifteen of the waning (moon)." [1]

The significance of stopping after the fourteenth operation is obvious. For after that there will be the repetition of the previous factors.

[1] *ŚBr*, x. 4. 2. 2-17. The translation is substantially due to Eggeling. We have only introduced the ciphers for the numbers in words.

APPENDIX

Some Technical Terms of the Śulba

Line of Symmetry of the Vedi.—Every one of the altars of various shapes that have been described in the *Śulba*, has a line of symmetry. Some which are square, rectangular or circular (with or without spokes) have, indeed, more than one such line. But primary importance is always attached even in those cases only to one of them. That line of symmetry of an altar is technically called the *pṛṣṭhyā*. This term is derived from the word *pṛṣṭha* (or "back") and so means "the line marking the back or rather the back-bone of the altar." It has its origin in the comparison of the altar with an animal which occurs repeatedly in the *Saṁhitā* and *Brāhmaṇa*. For instance, the *Taittirīya Saṁhitā* observes, "The Fire-altar is an animal."[1]

Configuration of the Vedi.—A sacrificial altar is built in such a configuration as to place its principal line of symmetry always along the west-to-east direction. Hence it is also called the *prācī*, or "the eastward line." This line, as has been already observed, is of primary importance in the geometry of the *Śulba*. For all constructions are described in the *Śulba* invariably with reference to it. The sides of an altar lying on either sides of its *prācī*, whether parallel to it or not, are called its *pārśvamānī*, from *pārśva*="side" and *māna*="measure," and hence meaning literally the "side measure;" those which are at right angles to the *prācī* are called the *tiryaṅmānī* or the

[1] *TS*, v. 2. 10. 1.

transverse measure," from *tiryak*="transverse," *māna*="measure." The latter term is oftentimes called in short *tiryak*, *tiraścīna* or *tiraścī* ("transverse") which are sometimes further abbreviated into *tiraḥ*. These terms are very old and occur in the earliest literatures of the Hindus.[1] The transverse sides are again distinguished into *paścāttiraścī* ("the western transverse side") and *purastāttiraścī* (" the eastern transverse side").[2] The former is also called the *mukha* (="the face") and the latter the *pada* (="the base") of the altar.

Line.—The line is called in the *Śulba*, *lekhā* or *rekhā*, both the terms being identical, as, according to the rules of the Sanskrit Grammar, the alphabets *l* and *r* can be replaced mutually. A straight line is distinguished as *ṛjulekhā*, *ṛju* meaning "straight."[3]

Rectilinear Figures.—In the *Śulba*, we discern two different systems of nomenclature for the rectilinear geometrical figures.[4] In one system the naming is according to the number of angles or corners in the figures, and the names are formed by the juxtaposition of the number names with *aśra* or *asra* which ordinarily means "corner," "angle," *e.g.*, *tryasra* (" triangle"), *caturasra* ("quadrangle"), etc. These names were introduced in the time of the *Śrauta-sūtra* (c. 2000-1500 B.C.). Still older names were compounds ending with *srakti* (= " angle," " corner "). Thus the name *catuḥsrakti*, which literally means, the " quadrangle," occurs in the *Vājasaneyī Saṁhitā*,

[1] For instance see *TS*, vi. 2.4.5 ; *MaiS*, iii. 8.4 ; *ŚBr*, vii. 1.1.18 ; etc.

[2] *BŚl*, i. 72. 76 ; *TS*, vi. 2.4.5.

[3] *BŚl*, ii. 32.

[4] For fuller information on this point, see the author's article, " On the Hindu Names for the Rectilinear Geometrical Figures," in the *Journal of the Asiatic Society of Bengal*, N.S., XXVI, 1930, pp. 283-290.

Taittirīya Saṁhitā, *Śatapatha Brāhmaṇa*, *Āpastamba Śrauta*, *Baudhāyana Śulba* and other works. In the *Ṛg-veda*, we find the term *navasrakti* referring to the "nine corners" of the heaven. In the *Kātyāyana Śulba Pariśiṣṭa*,[1] we have compound names for rectilinear figures ending with *karṇa*. The Sanskrit word *karṇa* means the "ear." Applied to geometrical figures, it implies the "angle"; hence *trikarṇa* = "triangle," *pañcakarṇa* = "pentagon." The word *karṇa* degenerated into *koṇa* in the *Prākṛta* languages. So in the Ardha Māgadhī work *Sūryaprajñapti*,[2] we get the names *trikoṇa* (= "trigonon"), *catuṣkoṇa* (= "tetragonon"), *pañcakoṇa* (= "pentagon") etc. These terms are, however, accepted in later Sanskrit literature.[3] The term *aśra* or *asra* in a compound name sometimes denotes the "side" Thus Baudhāyana once described a square as *catuḥsrakti* ("four-cornered") and *sama-caturasra* ("equi-four-sided").[4] So the terms *tryasra*, *caturasra*, etc., will also mean respectively "trilateral," "quadrilateral," etc.[5] Thus we get a second system of naming retilinear geometrical figures according to the number of sides they possess.

An isosceles triangle is denoted by the term *praüga*. This word is probably derived from *pra* + *yuga*, meaning "the forepart of the shafts of a chariot." A rhombus is similarly called *ubhayataḥ praüga* ("praüga on both sides") inasmuch as it is divided into two praügas by a diagonal. Both these terms are as old as the *Taittirīya Saṁhitā*[6]

1 *KŚlP*, iv. 7-8.

2 *Sūryaprajñapti*, Sūtra 19, 25.

3 See for instance, the *Pariśiṣṭas of the Atharva-veda*, xxiii. 1 5 xxv. 1, 3, 6, 7, etc. ; *Arthaśāstra of Kauṭilya*, ii. 11. 29.

4 *BŚl*, i. 79.

5 *Cf. ĀpŚ*, xx. 12.

6 *TS*, v. 4. 11.

and continued to be used in the same sense in posterior works, the *Brāhmaṇa* and *Śrauta* including *Śulba*.

A square is generally called *sama-caturasra* (*sama*= "equal"). It is oftentimes, of course when there is no chance of ambiguity the context being clear, shortened into *caturasra*,[1] and occasionally even into simple *sama*.[2] Thibaut is responsible for the opinion that in the term *sama-caturasra*, the word *sama* refers to the equality of four sides and *caturasra* implies that the four angles of the figure are right angles.[3] A more plausible interpretation would be that *sama* refers to the form or shape of the figure which is to be the same in every respect, *caturasra* implying a quadrangle or quadrilateral. It will then be consistent with the term *dīrgha-caturasra* for the rectangle,[4] which signifies that the form of the *caturasra* is in this case *dīrgha* (or "longish"). The rectangle is sometimes called in short the *dīrgha*.[5] The term for a quadrilateral of unequal sides is the *viṣama-caturasra* (literally, "inequilateral quadrilateral"). But in contradistinction from the *sama-caturasra* or the square, that term may denote also the rectangle.[6]

When all the angles of a polygon are equal, it is said to be of *eka-karṇa* (literally, "one-angled") variety; and when not so, of *dvi-karṇa* (literally, "two-angled") variety, implying that in this variety the angles of the figure are of more than one size).[7]

The diagonal is called the *akṣṇa* or *akṣṇayā* ("that which goes across or transversely," that is, "the cross-

[1] *BŚl*, i. 22. 50. 51 ; *ĀpŚl*, i. 5 ; ii. 4- 5 ; etc.

[2] *ĀpŚl*, i. 5,

[3] Thibaut, *Śulvasūtras*, p. 7.

[4] *BŚl*, i. 36. 38 ; *ĀpŚl*, ii. 7 , iii. 1 ; etc.

[5] *ĀpŚl*, i. 4.

[6] See *KŚl*, iii. 4.

[7] *Vide supra* p. 81, foot-note 2.

line")[1] In relation to the instrument of measurement, it is sometimes designated as the *akṣṇayā-rajju* ("the diagonal cord")[2] and at other times the *akṣṇayā-veṇu* ("the diagonal bamboo-rod").[3] The diagonal is also denoted by *karṇa*, meaning "the line going across the angle" or "the line going across from corner to corner."[4]

Circle.—In the *Śulba*, the circle is designated the *maṇḍala* ("round").[5] *pari-maṇḍala* ("round on all sides");[6] the circumference, *parināha*[7] ("bounding line on all sides"), and the diameter, *viṣkambha*[8] or *vyāsa* ("breadth"). The centre of the circle is called *madhya* ("middle")- But this term is also used in more general sense for the middlemost point of a square or rectangle,[9] or of a line.[10] The segment of the circle is denoted by the term *pradhi*.[11]

It is perhaps noteworthy that the direction of rotation was used to be indicated in the Vedic Age by means of an

[1] *BŚl*, i. 52; iii. 55. 65; *BŚr*, x. 19, xix. 1. Compare *ṚV*, viii. 7.35; *ĀpŚl*, ii. 5.

Rāma, the commentator of the *Kātyāyana Śulba*, is of opinion that this line is so called because it divides the figure into two *akṣi* or "eyes";

"अक्षिवत् नयति चतुरस्रमक्ष्णया ··· चतुरस्रं कोणात् प्रतिकोणं नीता हि मध्यरज्जुश्चतुरस्रमक्षिद्वयरूपतां नयति । तथा अस्मिन्नपि तिर्य्यङ्मानीपार्श्वमान्यन्तौ नीताक्ष्णया भवति ।" *KŚl*, ii. 7 (Com.)

[2] *BŚl*, i. 50, *ĀpŚl*, i. 325.

[3] *ĀpŚl*, ix. 3.

[4] *BŚr*, xix. 1; *KŚr*, xvii. 6. 3.

[5] *BŚl*, i. 23, 24, 58, 59; *ĀpŚl*, iii. 2, 3.

[6] *ĀpŚl*, vii. 6. 13; *BŚl*, ii. 63. 70.

[7] *BŚl*, i. 113.

[8] *BŚl*, i. 23, 25, 26; *ĀpŚl*, iii; *ĀpŚl*, vii. 10.

[9] *BŚl*, i. 58; *ĀŚl*. iii. 2.

[10] *BŚl*, i. 56, 57, 73; *ĀpŚl*, ii. 1.

[11] *BŚl*, ii. 71-2.

arc of a circle very likely with an arrow-head at one extremity. Thus we have the terms *dakṣiṇāvṛta lekhā* ("the line turning rotationally towards the right ") and *savyāvṛta lekhā* ("the line turning rotationally towards the left").[1] Again a rotation is called *dakṣiṇā-prāk*, if it is towards the east by the south, and *dakṣiṇā-pratyak*, if it is towards the west by the south.[2]

Area.—In the early Hindu geometry a figure is generally denoted by the term *kṣetra*[3] and its area by *bhūmi*.[4] Occasionally, however, the term *kṣetra* is employed also in the sense of an area.[4]

Fundamental Operations.—Addition is called *samāsa* ("putting together") and the sum obtained *samasta* (" whole," "total").[5] Subtraction is called *nirhāra* ("deduction") and the remainder *śeṣa*.[6] Division is *bhāga*, *vibhāga*. One term deserves special notice : it is *abhyāsa*. This word, formed from *abhi* and *āsa*, means radically "repetition," "reduplication." It then came to denote, in its various declinations, the operation of addition[7] as well as of multiplication.[8] Whence it seems that the early Hindus recognised multiplication to be a kind of addition.

[1] *BŚl*, ii. 30-31.
[2] *ĀpŚl*, viii. 9-10.
[3] *BŚr*, xix. 7-9.
[4] *BŚr*, xxvi. 25 ; *BŚl*, i. 56, 57, 82 ; *ĀpŚl*, i. 5. *KŚl*. iii. 11.
[5] See *BŚl*, i. 59 ; *ĀpŚl*, ii. 4.
[6] *BŚl*, i, 58 ; *ĀpŚl*, ii. 6. 7.
[7] *BŚl*, ii. 4. 9. 11 ; *ĀpŚl*, i. 9, 2 ; ii. 1.
[8] *ĀpŚl*, v. 3.

BIBLIOGRAPHY OF THE SULBA

I

The following works on the *Sulba* and their commentaries, published and unpublished, have been made use of in this book.[1]

1. *Baudhāyana Sulba* :

This work was published by G. Thibaut with English translation, critical notes, extracts from the commentary of Dvārakanātha Yajvā and a few diagrams, in the *Paṇḍit*, a monthly journal, now defunct, of the Sanskrit College of Benares : Old Series, IX and X, 1874-5 ; New Series, I, 1877. The text is printed in the *Baudhāyana Srauta Sūtra*, being the 30th Chapter of it, edited by W. Caland,

[1] In the collection of the Calcutta University, there is a good number of transcripts of the works on the *Sulba* and their commentaries. We have noted below, along with them, the original manuscripts from which they have been transcribed. The key to the abbreviations used is as follows :

Ady. Lib. = Library of the Theosophical Society at Adyar, Southern India.

Asiat. Soc. Ben. = Asiatic Society of Bengal, Calcutta.

Bhand. O. Inst. = Bhandarkar Oriental Institute, Poona.

Bom. Br. Roy. Asiat. Soc. = Bombay Branch of the Royal Asiatic Society, Bombay.

Bom. Univ. = Bombay University.

Ind. Off. Lib. = Library of the India Office at London.

Mad. O. Ms. Lib. = Government Oriental Manuscripts Library, Madras.

Mys. O. Ms. Lib. = Government Oriental Manuscripts Library, Mysore.

Tanj. Pal. Lib. = Palace Library of the Mahāraja of Tanjore.

in three volumes, Calcutta, 1904, 1907, 1913. Besides these printed works, the manuscripts consulted are:

(1·1) Text only:

C. U. 21

(1·2) Text with the commentary, entitled *Śulba-dīpikā*, of Dvārakanātha Yajvā:

C. U. 15 (=Tanj. Pal. Lib., No. 3743B).

C. U. 15 (*a*) (=Tanj. Pal. Lib., No. 3742, collated with No. 3743B).

C. U. 21 (=Tanj. Pal. Lib., No. 9160(*b*), collated with No. 3743B),

(1·3) Text with the commentary, *Śulba-mīmāṁsā*, of Veṅkateśvara Dīkṣita:

C. U. 42 (=Bhand. O. Inst., No. 96 of 1891-95).

2. *Āpastamba Śulba*:

This has been edited and published by A. Bürk, with German translation, notes and comments, helpful extracts from the commentaries of Kapardisvāmī, Karavindasvāmī and Sundararāja, and diagrams, together with a masterly introduction, in the *Zeitschrift der deutschen morgenländischen Gesellschaft*, LV, 1902, pp. 543-591 ; LVI, 1903, pp. 327-391. Manuscripts consulted are:

(2·1) Text only:

C. U. 16 (=Tanj. Pal. Lib., No. 3846).

C. U. 27 (=Bom. Univ.).

C. U. 34 (=Ady. Lib.).

(2·2) Text with the commentary, *Śulba-vyākhyā*, of Kapardisvāmī:

C. U. 1 (=Mad. O. Ms. Lib., No. 777 collated with No. 151 (*b*)).

C. U. 17 (=Tanj. Pal. Lib., No. 3851).
C. U. 52 (=Mad. O. Ms. Lib.).
C. U. 54 (=Mys. O. Ms. Lib.).

(2.3) Text with the commentary, *Śulba-pradīpikā*, of Karavindasvāmī:
C. U. 18 (=Tanj. Pal. Lib., No. 3852),
C. U. 31 (=Ms. of Pandit Gopal Desikacarya of Kumbhakonam).
C. U. 49 (=Mad. O. Ms. Lib.).
C. U. 50 (=Do. another copy).
C. U. 53 (=Mys. O. Ms. Lib.),

(2·4) Text with the commentary, *Śulba-pradīpa*, of Sundararāja :
C. U. 10 (=Ms. of Pandit Gopal Desikacarya of Kumbhakonam).
C. U. 19 (=Tanj. Pal. Lib., No. 9160 (*a*)).
C. U. 20 (=Tanj. Pal. Lib., No. 9160 (*a*) collated with 9172 (*a*)).
C. U. 32 (=Ady. Lib.).
C. U. 33 (=Do., another copy).
C. U. 43 (=Bhand. O. Lib., No. 23 of A 1879-80).
C. U. 44 (=Mad. O. Ms. Lib. No. R 911 (*a*).
C. U. 51 (=Mad. O. Ms. Lib., No. 5812).
C. U. 55 (=Mys. O. Ms. Lib.).

(2·5) Text with the commentary, *Āpastambīya Śulba-bhāṣya*, of Gopāl.

(3) *Kātyāyana Śulba*:

A portion of this work, only the first two chapters, was published by G. Thibaut with English translation and the commentary, entitled the *Śulba-sūtra-vṛtti* of Rāma (*c*. 1450 A.D.) in the *Paṇḍit*, New Series, IV, 1882. Manuscripts consulted are:

(3.1) Text only :

C. U. 41 (=Bhand. O. Inst., No 74 of A 1881-82). It contains also the transcript of the MS. of the *Pariśiṣṭa* of *Kātyāyana Śulba* in the Ind. Off. Lib., No. E 363.

C. U. 45 (=Bom. Br. Roy. Asiat. Soc. Lib.).

(3·2) Text with the commentary, *Śulba-Sūtra-vivaraṇa*, of Mahīdhara (1589 A.D.):

C. U. 41 (=Bhand. O. Inst., No. 363 of 1883-84).

C. U. 46 (=Bom. Br. Roy. Asiat. Soc. Lib.).

4. *Mānava Śulba*:

C. U. 23 (=Asiat. Soc. Ben., No. I. E. 17); the text with the commentary of Śivadāsa.

C. U. 29 (=Bom. Univ.); the text only.

5. *Maitrāyaṇīya Śulba*:

C. U. 47 (=Bom. Br. Roy. Asiat. Soc.); the text with the commentary of Śaṅkara.

6. *Varāha Śulba*:

C. U. 30 (=Mys. O. Ms. Lib.).

II

Following is a list of articles dealing primarily with the mathematics in the *Śulba*:

(1) Bürk, A.—"Das Āpastamba Sulva-sūtra," *Zeitschrift der deutschen morgenländischen Gesellschaft*, LV, 1901, pp. 543-591; LVI, 1902, pp. 327-391.

(2) Cantor, M.—"Über die älteste indische Mathematik," *Archiv der Mathematik und Physik*, VIII (3), 1905, pp. 63-72.

(3) ,, — "Gräko-indische Studien," *Zeitschrift für Mathematik und Physik*, XXII, Hist. Lit.

(4) Datta, B—"The Origin of Hindu Indeterminate Analysis," *Archion*, XIII, 1931, pp. 401-7.

(5) Levi, B.—" Observazioni e congetture sopra la geometria degli Indiani," *Bibliotheca Mathematica*, IX (3), 1909/10, p. 97.

(6) Mazumdar, N. K.—" Mānava Śulba Sūtram," *Journal of the Department of Letters in the Calcutta University*, VIII, 1922.

(7) ,,—"On the Different Śulba Sutras," *Proceedings and Transactions of the Second Oriental Conference*, Calcutta, 1922, pp. 561-4.

(8) Milhaud, G.—" La geometric d'Āpastamba," *Revue générale des Sciences*, XXI, 1910, pp. 512-520.

(9) Müller, C.—" Die Mathematik der Śulvasūtra," *Abhund. a. d. Math. Sem. d. Hambung. Univ.*, Bd. VII. 1929, pp. 173-204.

(10) Thibaut, G.—" On the Sulva-sūtras," *Journal of the Asiatic Society of Bengal*, XLIV, 1875, pp. 227-275.

(11) ,, —" The Baudhāyana Śulva-sūtra. *The Paṇḍit*, *Old Series*, IX, X; New Series, I.

(12) ,, —" The Kātyāyana Śulva-sūtra; " *The Paṇḍit*, New Series, IV.

(13) Vogt, H.—" Der Pythagoreischen Lehrsatz in der ältesten Geometrie der Inder," *Schleische Gesellschaft, Jahresberichte der Math. Sec*, LXXXIV, 1906, pp. 3-4.

(14) ,, —" Haben die alten Inder den Pythagorsischen Lehrsatz und das Irrationale gekant ?" *Bibliotheca Mathematica*, VII. (3), 1906/7, pp. 6-23.

(15) Weber, A.—" Zur Kenntniss des vedischen Opferrituals," *Indische Studien*, XIII, pp. 216-292.

(16) Zeuthen, H. G.—" Sur l'Arithmétique Géometrique des Grecs et des Indiens," *Bibliotheca Mathematica*, 1904, pp. 97-112.

(17) Zeuthen, H. G.—" Theorème dé Pythagore, Origine de la Geometrie scientifique," *Comptes Rendus du IIme Congrés internationale de Philosophie*, Genéve, 1904.

III

Valuable information on the subject can also be had from the following standard works and articles:

(1) Cantor, M.—*Geschichte der Mathematik*, I_3, pp. 635-645.

(2) Datta, B.—" Hindu Contribution to Mathematics," *Bulletin of the Mathematical Association of the University of Allahabad*, I to II, pp. 1ff.

(3 ,, —" The Scope and Development of the Hindu *Gaṇita*," *Indian Historical Quarterly*, V, 1929, pp. 479-512.

(4) Dutt, R. C.—*A History of Civilisation in Ancient India*, revised edition, London 1893, I, pp. 260 ff.

5) Hankel, H.—*Zur Geschichte der Mathematik im Alterthum und Mittelalter*, Leipzig, 1874.

(6) Heath, T.—*The Thirteen Books of Euclid's Elements*, I, Cambridge, 1908, pp. 352-364.

(7) Von Schroeder, L.—*Indien Literatur und Kultur*, Leipzig, 1887.

(8) ,, —*Pythagoras und die Inder*, Leipzig, 1884.

INDEX